The Fibonacci Pulse

A Practical Guide to Soil Intelligence

ZEN KNEZEVIC

ISBN — eBook: [to be assigned]

ISBN — Paperback: [to be assigned]

ISBN — Hardcover: [to be assigned]

First published in 2026.

Disclaimer

This book is intended for informational and educational purposes only. It reflects the author's perspectives, interpretations, and experience regarding soil health, regenerative agriculture, and ecological systems.

The content of this book is not intended as, and should not be construed as, professional agricultural, agronomic, environmental, medical, legal, or financial advice. Readers are encouraged to consult appropriately qualified professionals before implementing any practices or making decisions that may affect land, crops, health, safety, or livelihood.

The author and publisher make no representations or warranties regarding the accuracy, completeness, or applicability of the information contained herein and disclaim any liability for loss, damage, or injury arising from the use or application of the ideas, methods, or practices described.

All examples and scenarios are illustrative. Outcomes may vary depending on environmental conditions, geographic location, management practices, and other factors beyond the author's control.

Responsibility for the interpretation and application of this material rests solely with the reader.

**

For **Leonardo Fibonacci**,
who gave shape to the spiral...

For the countless **ancestors**,
named and unnamed,
who tended the living world before us
and left more than they took...
For the **Seventh Generation**...

For the **soil** that remembers us
long after we are gone.

**

"We shape the land for a time, but the land shapes us forever."

The Vernaen

Table of Contents

Prologue

Beneath the Surface

Most of what determines the fate of a landscape happens where we are not looking.

Beneath the boots, beneath the plough, beneath the data sheets and yield maps, a living architecture is quietly at work. It is not inert. It is not passive. It is patterned, responsive, and intelligent in ways modern management has largely forgotten how to see.

One teaspoon of healthy soil contains more living organisms than there are people on Earth. Bacteria, fungi, protozoa, nematodes—an entire civilisation compressed into a space small enough to rest on the tip of a finger. This is not a metaphor. It is a measurable reality, and it should fundamentally change how we think about control, intervention, and stewardship.

For generations, we have been taught to look at soil as a medium—something to be corrected, supplemented, pushed, or fixed. A substrate that responds predictably to inputs. A problem to be solved one deficiency at a time. This way of seeing has produced impressive short term gains, but it has also trained us to ignore the deeper system beneath our feet: the structure that holds air and water, the networks that move minerals and information, the rhythms that govern recovery, resilience, and collapse.

Soil, like any living system, does not behave linearly. It does not improve by addition alone. It compounds—or it degrades—based on whether its internal relationships are allowed to form, connect, and reinforce one another. When those relationships are intact, the system builds upon itself. When they are repeatedly interrupted, no amount of external correction can restore what has been lost.

This book was not written to offer another set of techniques layered on top of an exhausted model. It was written to name a deeper operating system—one that has always been there, quietly shaping outcomes regardless of our intentions. The same recursive patterns that appear in forests, coastlines, circulatory systems, and galaxies also govern the underground world of roots, fungi, aggregates, and pore space. These patterns are not metaphors. They are instructions.

To work with soil effectively is not to dominate it, nor to leave it alone entirely. It is to recognise where intelligence already exists and to stop interfering with the conditions it requires to express itself. This demands a different role for the human steward—not as controller or fixer, but as observer, protector, and conductor of processes that operate on timescales longer than a season or a balance sheet.

What follows is an invitation to look differently. To measure differently. To slow down long enough to notice where systems are compounding and where they are being reset to zero. The practices described here are practical and measurable, but they rest on a simple premise: that soil remembers what we do to it, and responds accordingly.

The pulse is already there.

The question is whether we are working with it—or against it.

The Waking of the Watcher

The Waking of the Watcher

"I awoke to the scream of iron's bite,
Where once the deep stone softly hummed its right.
The constellations' sacred math is torn,
And from the living bone, the marrow shorn.
I rise from barren dust, the linear reign,
To wake the ancient spiral once again."

Introduction

Welcome to the Pulse

Something is wrong with the way we've been taught to see the earth.

You probably already feel it. You've done the reading. You've planted the cover crops. You've cut back on the chemicals, switched to no till, watched regenerative farming videos at midnight on your phone. You've put in the work. And yet the soil still cracks in July. The yields still disappoint. The bank balance still tightens your chest. You're doing everything the manuals say—and it still feels like you're pushing a boulder uphill in the dark.

That feeling has a name.

It's called the **Linear Void**.

The Linear Void is the invisible operating system of modern agriculture. It's not a place you visit—it's the lens you were handed early on. The lens that says the world is made of separate parts. That soil is a substrate. That growth comes from adding inputs. That efficiency is virtue. That the only number that matter is the one at the bottom of a profit statement.

It's the math of *1 + 1 = 2*. Clean. Predictable. And completely wrong for anything that breathes.

Here's what that mindset has actually produced: compacted fields that shed water like asphalt. Topsoil thin enough to read a newspaper through. Farms that require more fertilizer every year just to stand still. Farmers who work harder for less, season after season, wondering what they're missing.

The Linear Void doesn't just exhaust land.

It exhausts people.

But the exhaustion isn't a personal failure.

It's a **frequency mismatch**.

Nature does not move in straight lines. It never has. Look at the spiral of a sunflower head. The unfurling of a fern. The branching of a river delta. The curve of a hurricane seen from space. These forms aren't decorative accidents—they are expressions of a recursive pattern that builds by remembering what came before.

That pattern is the Fibonacci sequence.

1, 1, 2, 3, 5, 8, 13.

Each number carries the memory of the previous two. Growth that compounds rather than resets. Expansion without exhaustion. A system that strengthens because it never works against itself.

This is the **Fibonacci Pulse**.

And your soil already knows it.

Before we go any further, it's important to be clear about what this book is—and what it is not. Everything that follows is accountable to what can be observed on land. If an idea cannot be seen in soil structure, measured in water movement, or felt underfoot after rain, it does not belong here. You will not be asked to believe. You will be asked to look, to test, and to notice what repeats across seasons and scales. The language may stretch at times—living systems require a broader vocabulary than spreadsheets allow—but the standard remains simple: what works must be visible, repeatable, and answerable to the land itself.

The problem isn't that the earth has forgotten how to function.

The problem is that we keep interrupting the rhythm it runs on.

Every time a plough tears through the upper soil, a biological network decade in the making is shredded in minutes. Every time land is left bare through winter, a solar pump that feeds billions of organisms is switched off. Every time fertility is purchased instead of regenerated, we choose a metronome over a melody.

This book is not another manual layered on top of that exhaustion.

This book is the music.

Think of it as a **tuning fork**. Not a checklist. Not a laminated set of steps. A tuning fork, because what's missing isn't more information—it's resonance. A way of working that aligns the mud on your boots with the deeper physics that govern living systems everywhere.

If you read this book looking for recipes, you may miss it.
If you read it looking for relationships, it will change the way you see land forever.

Consider someone like *Vernaen*—a composite of the farmers who inspired this work. He's in his forties, farming a few hundred acres in a dry inland valley. He's been chemical free for years. He plants diverse covers. Rotates stock carefully. Reads everything Gabe Brown publishes. And yet when a soil core is pulled, the aggregates still crumble. Water still pools after rain. Vernaen knows the *what*. He's been doing the *what* faithfully. What he doesn't yet have is the *why*—the architecture that allows a system to carry itself forward.

This book is written for Vernaen.
And it's written for you.

The shift we're talking about isn't merely agricultural—it's structural. It's the move from fixing parts in isolation to conducting a living system. From extraction to stewardship. From reacting to symptoms to restoring signal.

That word—**Observer**—will matter more than it seems right now. Later chapters will explore what biology and physics actually say about the role of presence in living systems. For now, it's enough to know this: the way you see your land changes what your land can become. The Fibonacci Pulse is not just a pattern—it's a relationship. And relationships only work when both parties are paying attention.

Throughout this book, we'll return to a single metaphor.

A cathedral.

If you walk into Notre Dame, the most important part of the building isn't the stone. It's the air. The structure exists to protect a volume of space where something sacred can happen. Soil works the same way. Minerals are the stone. The real cathedral is the pore space—the invisible architecture of air and water channels held together by fungal glues like glomalin. When that space collapses, the building still stands. But the spirit has gone.

The Fibonacci Pulse: Activating the Source Code of the Soil is a guide to rebuilding that cathedral—aggregate by aggregate, layer by layer—from the microscopic to the generational.

What You Will Discover

This book rests on a single claim: when stewardship aligns with recursive natural rhythms, soil stops being a burden and becomes a self multiplying engine.

You'll learn how to calculate a **Soil Resonance Score (SRS)**—a framework grounded in Fibonacci weighting that shifts measurement away from extraction and toward construction. Architecture, Armor, Diversity, Biology, and Chemistry are weighted according to their foundational importance. The inversion is intentional. The most fundamental layers matter most.

You'll learn the **Slake Test**, a five minute ritual that tells you more about soil integrity than most lab reports. You'll understand the **Mycelial Internet**, the underground economy that has operated for four hundred million years. You'll learn the **Circadian Inhale**, the nightly breathing cycle of soil—and why timing matters more than force.

You'll also encounter ideas most manuals avoid—not as mysticism, but as physics applied to life. Observation. Presence. Responsibility across generations. By Chapter 9, these ideas will no longer feel abstract.

What makes this book different is the bridge. Most texts offer practice without theory, or philosophy without application. This book insists on both. The glomalin and the golden ratio. The slake test and the spiral. Mud on your boots, stars in your eyes.

The Roadmap to Resonance

The fifteen chapters that follow trace a spiral, not a straight line. Each builds on what came before, compounding understanding the way living systems do.

You'll begin underground, with structure. Move through pattern, timing, connection, protection, diversity, and exchange. You'll confront the Linear Void directly. And you'll emerge with a new identity—not as a manager of land, but as a steward embedded in time.

Each chapter ends with **Field Protocols**—specific, observable actions you can take within seven days. These are not suggestions. They are calibrations. By the end, you won't just understand the Fibonacci Pulse.

You'll feel it under your feet.

The soil doesn't need you to work harder.
It needs you to stop working against it.

The pulse has been there for four hundred million years.
You are not creating it.

You are tuning back in.

Chapter 1: The Silent Architecture

The Ancestor Speaks:

"Child of my blood, close your eyes. No, not to darkness, but to memory. Reach back through the seasons, through the generations, until you find the warmth of my hand, guiding yours. It is early morning. The sun, a soft blush on the horizon, has not yet kissed the dew from the leaves. We are standing where the wild strawberries hide, just beyond the last row of corn. Can you feel the cool earth beneath your bare feet?

"Now, listen. Not with your ears, but with your soul. Do you hear it? The silent breathing. This, my child, is the Earth's own inhale and exhale, the very pulse of life itself. My hand, weathered by countless suns and rains, reaches into the soil. I cup a handful, dark and rich, crumbling it gently between my fingers. 'Look,' I whisper, 'not at the dirt, but *into* it.'

"Do you see the tiny castles, strong yet delicate, that hold the world together? This is the work of the **Great Weaver**, tirelessly spinning threads of magic, binding the sand and silt into a sacred masonry that shelters the air, holds the precious water. Every crevice, every tiny pore, is a passageway, an invisible cathedral hallway where the spirit of the rain can linger, and the warmth of the sun can be drawn deep within. This **Silent Architecture** is the Earth's own lungs, her very breath depending on its intricate design.

"I remember the shame in our grandfather's eyes, generations ago, when the new tools came. The steel that tore the ground, making it bleed. The great machines that crushed these castles, silencing the breath. He saw the future in the dust, the **Linear Void** marching across the fields, leaving behind only fractured fragments. He knew, with a pang in his heart, that when the water could no longer linger, when the air could no longer flow, the Earth's voice would dim, and the **Fibonacci Pulse** of life would falter.

"He would take a dry clod from the broken fields and drop it into a cup of clear water—the 'slake test,' they call it now. Watch it crumble, almost instantly, a sigh of defeat. But then, he would hold another, dark and whole, from a patch untouched, and drop it. See how it holds its form, slowly softening, yet remaining whole, its tiny castles intact. 'This,' he would murmur, 'this is the truth spoken by the land itself. A judgment of the hand, yes, but a promise too.'

"Now, open your eyes. Hold your own hand before you. Do you feel the echo of that memory?

Can you almost touch the invisible architecture, hear the silent breathing? The **Seventh Generation** waits for you to remember, to rebuild. Will you listen to the Earth's breath, or will you let it suffocate?

The choice, as always, is held in your own hands."

Prelude: On rooms, breath, and what holds

The Silent Architecture

"No grave lies here, but air's grand cathedral,
Woven by the Weaver, silver, resilient.
Honor the earth, but guard each holy pore,
For where the water lingers, life draws breath.
The slake of the aggregate, a judgment of the hand,
Reveals the spirit's truth across the land."

The Silent Architecture: Rebuilding the Cathedral

Before anything else in this book, nothing will be asked of you except to look.

Walk into any great cathedral and stop moving for a moment. Don't look at the stone. Don't look at the glass. Look up—and then look at nothing at all. Look at the air.

That vast, breathing volume of space between the floor and the vaulted ceiling is the whole point. The stone exists only to protect it. The pillars exist only to hold it open. If you poured concrete from floor to ceiling and filled every inch of that space, the building would still be standing. But the cathedral would be dead. It would be nothing more than a block of rock with a name.

Your soil works the same way.

Most of what makes soil function is not solid. It is space. Space for air to move. Space for water to wait. Space for roots to explore and microbes to breathe. When that space collapses, the soil doesn't disappear—but its intelligence does.

This chapter is about learning to see that invisible architecture.

Until you can see it, every other conversation about fertility, yield, biology, or resilience is happening too late.

Seeing the Building, Not the Bricks

Modern agriculture taught us to think in parts. Nitrogen. Phosphorus. Potassium. The familiar trinity of N P K. The logic is straightforward: identify what's missing, add it back, get a result. It's the mindset of a mechanic replacing components one at a time.

For a while, that approach worked. Yields rose. Food became abundant. The system looked like progress.

But a mechanic working on individual parts cannot see what is happening to the structure as a whole.

This parts based way of thinking treats soil as a pile of materials to be managed rather than a system to be protected. We'll call this the **Brick Model**. In the Brick Model, the particles themselves are the story. Adjust the chemistry of the bricks and the system will behave.

The problem is simple: bricks piled together do not make a building.

A building requires structure. It requires the deliberate arrangement of solid material around empty space. A bedroom without air is not a bedroom. A lung without air is not a lung. And soil without pore space is not soil—it is dirt.

Pore space is the silent architecture. It is the network of microscopic channels and chambers between soil particles that allows life to move. In a healthy soil, roughly half the volume is solid and half is space. That space is not empty. It is filled with air, water, dissolved gases, fungal threads, and biological activity. It is the most productive real estate on your land, and it is almost never discussed.

When soil is disturbed repeatedly, that space collapses.

The structure may still look intact from above, but underground the rooms have caved in. Water can no longer enter easily. Air cannot circulate. Roots struggle. Biology suffocates. And the system begins to fail—not because nutrients are missing, but because the architecture that makes nutrients usable is gone.

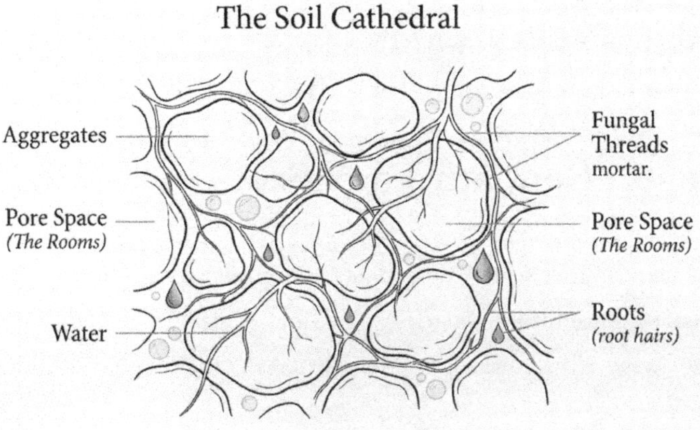

Figure 1.1. Soil Architecture: A cathedral built of air and water, held together by biological glue.

What Collapse Looks Like

Imagine pulling a clod of soil from the ground. Not a loose handful, but a firm piece that holds together in your palm. In healthy soil, that clod is made of aggregates—small, irregular shapes that lean against one another and create open space between them. Think of these aggregates as arched doorways. Their curves prevent them from packing tightly, leaving room for air and water to move.

When those aggregates are crushed by mechanical force, they shatter into flatter fragments. The curves are lost. The doorways collapse. The fragments pack together. The space disappears.

Then it rains.

With no open channels to enter, water sheets across the surface, carrying the finest and most biologically active particles with it. This is erosion—but more precisely, it is architectural failure. The system can no longer do what soil evolved to do: hold water, breathe air, and sustain life.

Many farmers experience this without ever being told what they are seeing.

Consider a farmer we'll call **Daniel**. He manages a few hundred acres in a semi arid region and has watched his topsoil thin for years. He's followed the recommendations. He's applied amendments. He's invested in inputs. And yet after heavy rain, water still runs off his fields in brown sheets—while an undisturbed pasture nearby absorbs the same rain without losing a grain.

Daniel doesn't lack knowledge. He lacks rooms.

The Mason Beneath Your Feet

If soil is a cathedral, something must be holding it together.

For most of agricultural history, we knew healthy soil had structure, but we didn't know why. We could see the difference. We just couldn't name the mechanism.

In the 1990s, soil scientists identified it.

A carbon rich glycoprotein coats the surfaces of soil particles in healthy systems, binding them into stable aggregates. That substance is called **glomalin**. It acts as a biological mortar, holding the irregular shapes together and keeping pore space open under pressure.

Glomalin is not added to soil. It is produced.

Specifically, it is produced by arbuscular mycorrhizal fungi as they grow through the soil. As fungal threads extend, they coat themselves in glomalin. When those threads die back, the glomalin remains, reinforcing structure and building resilience over time.

The fungi are the masons.
Glomalin is the mortar.
The pore space is the cathedral.

Soil rich in glomalin can hold many times its weight in water. Not because it is fluffy, but because it is structured. This is the difference between soil that survives drought and soil that cracks under it.

Here is the critical point: you cannot manufacture glomalin. You cannot purchase it. You cannot apply it. You can only create the conditions that allow the masons to work.

Repeated disturbance destroys both the building and the builders. Tillage shatters aggregates and severs fungal networks in a single pass. The soil may look loose for a short time, but the structure is gone. When rain arrives, the clods collapse and compaction returns—often worse than before.

Only biology builds architecture that lasts.

The Slake Test: Let the Soil Speak

You do not need a laboratory to know whether your cathedral is standing.

You need a jar of water, a clod of soil, and a few minutes of attention.

This is the **Slake Test**, and it is the most honest conversation you can have with your land.

Take a clod of soil from a representative area—not the best corner, not the worst. Let it air dry until it is brittle. Fill a clear jar with water. Suspend the clod in the water and watch.

If the clod melts quickly, releasing clouds of silt, the architecture is gone. There is no mortar. When rain hits that soil, it will not enter—it will run.

If the clod holds its shape, dark and intact beneath the water, the structure is present. The pore space remains open. Rain will enter. Life will persist.

Do this test in more than one place. Compare a managed field with a fence line or undisturbed edge. The contrast is often unmistakable. Same soil type. Same climate. Different architecture.

This is not theory. This is evidence.

Record what you see. Photograph it. Date it. This is your baseline. You cannot rebuild what you have not first measured.

Protecting the Pore

Everything that follows in this book depends on what you've seen here.

Without architecture, nothing compounds. Without pore space, biology suffocates. Without structure, every other effort multiplies against collapse.

Your first responsibility as a steward is simple: **protect the pore**.

Before disturbing soil, ask one question: *Is this action preserving structure, or demolishing it?* If the answer is unclear, the safest choice is restraint.

Over the next two weeks:

1. **Run the Slake Test** in at least three locations and record the results.
2. **Walk your fields after rain**, noticing where water enters and where it runs.
3. **Identify one area** you will leave undisturbed for twelve months as a reference point.

These are not symbolic gestures. They are calibrations.

The soil beneath your feet has been building structure for hundreds of millions of years. The masons are already there. The mortar is ready to be made.

Your work, for now, is not to fix.

It is to stop breaking.

In the next chapter, we will look at the mathematics that govern how systems rebuild once interruption ends—and why compounding, not effort, determines the speed of recovery.

Before there can be rhythm, there must be rooms.

Chapter 2: The Math of Grace: The Fibonacci Sequence in the Field

The Ancestor Speaks:

"My dearest one, come closer now. The air is warm, thick with the scent of blossoming elderflower. We sit here, at the edge of the ancient oak, its roots like wise, gnarled fingers gripping the earth. Look up, into the perfect spiral of the sunflower, its countless seeds arranged in dizzying perfection. My Grandmother, your great-great-grandmother, showed me this very pattern, her eyes alight with a secret joy. 'This,' she would say, 'this is the true language of the Creator. The Math of Grace.'

"We were taught a different counting then, one that was all straight lines and hard edges. One seed in, one grain out. Two sheep sheared, two fleeces sold. The 'math of the tired man,' she called it, this ceaseless adding, this constant measuring of what we take. It makes us weary, doesn't it? Always chasing the balance, always feeling a lack. It is the Linear Void disguised as progress, promising abundance but delivering only exhaustion. Can you feel that old ache in your bones, that constant striving, never quite enough?

"But look again at the sunflower. Count the spirals, child. Find the sequence. One, one, two, three, five, eight... Do you see? It is not just addition; it is something more. Every new seed born from the last, yet each one creating the space for more, infinitely more. This is the Fibonacci Pulse, the very Source Code of the universe, echoing from the smallest fern frond to the grandest galaxy. It is the 'Spirit of the Spiral' revealing itself.

"My grandmother did not speak of 'golden ratios' or 'SRS scores' in numbers. She spoke of feeling the land's song. When the soil was truly alive, when the water sang through its hidden channels and the roots twined like lovers, she said you could feel the compounding. A single seed, tended with love and understanding, became not just a plant, but a whole ecosystem of abundance. It wasn't just 'addition' of plants, it was a 'compounding' of life, of resilience, of fertility that grew upon itself, year after year, more than you could ever simply count. It was the land returning more than was ever put in. Can you feel the shift from constant effort to effortless growth, the quiet confidence that comes from working with this deep wisdom?

"She would teach me to walk the fields, not with a ledger in my hand, but with my heart open. To feel the vibrancy beneath my feet, the quiet hum of growth. 'When the earth is singing its true song,' she would whisper, 'it hums in this spiral. It is the rhythm of generosity, of infinite return.' She knew that when we align our hands, our intentions, our very presence, with this Pulse, we stop being mere takers. We become Conductors. We help the Earth sing its most abundant song.

"Now, close your eyes. Can you see the spiral in your mind's eye? Can you feel its gentle, expansive rhythm? The Seventh Generation is not waiting for your additions; they are waiting for your compounding. They yearn for a future built not on what we take, but on the exponential grace of what we allow to flourish. Feel that expansion in your own heart. It is the truest math there is. It is the relief of knowing that abundance is the Earth's natural state, if only we listen."

Reflection: Pattern as memory

The Math of Grace

"The tired mind counts one and one again,
But the shell's coiled grace, the Master's sovereign plan.
Follow the ratio, 1.618—dirt's deep gold,
Till the SRS rings out, a story to be told.
Shed the flat lie; embrace the recursive truth."

The Math of Grace: The Fibonacci Sequence in the Field

You've seen the cathedral now.

You've held a clod of soil that either stayed whole or dissolved into ruin. You've watched architecture reveal itself in a jar of water—without a consultant, without a lab, without an argument. You have a baseline.

So why does doing the right thing still feel slow?

Why does progress sometimes seem to appear for a season, then vanish again—like the system forgets?

The answer isn't willpower. It isn't effort. It isn't that you haven't read enough, tried enough, or spent enough.

It's **math**.

Not the math you were taught at school. Not the math of spreadsheets and input budgets. But the math that living systems actually run on—the math that determines whether a change accumulates, or whether it resets back toward zero every year.

Most regenerative practitioners don't fail because they choose the wrong actions.

They fail because they apply the right actions in the wrong order, measure them with the wrong ruler, and then lose heart before the system has had time to compound.

This chapter is the ruler.

The Tired Math

Most of us were handed a single model for how the world works:

Add something, get something.

Add fertiliser, increase yield. Add sprays, reduce pressure. Add hours, increase output. Add more effort, more intervention, more inputs—because that's what the straight line demands when the results aren't coming fast enough.

It's clean. It's logical. And it produces an immediate sense of control.

It also produces diminishing returns.

Every farmer knows this feeling, even if they don't call it that. The first year, you get a response. The second year, the response shrinks. By the fifth year, you're applying more to hold the same result, and the system starts asking for additional products to compensate for the side effects of the first ones.

You're not managing a farm anymore. You're managing a dependency.

That is the Linear Void expressed as mathematics: a world where progress is measured in additions and subtractions, and the system becomes exhausted because it is forced to live against its nature.

Living systems do not grow by addition.

They grow by **recursion**.

Recursion: How Living Systems Remember

Recursion means a system uses its own recent past as the input for its next step. It doesn't start fresh. It doesn't reset. It carries memory forward.

Each output becomes part of the next input.

This is how forests build themselves without fertiliser. It's how a coral reef becomes an ecosystem rather than a pile of calcium. It's how soil becomes a living engine rather than a substrate.

You already know recursion without naming it.

You see it when a paddock improves slowly for years, and then suddenly seems to change character—holding moisture longer, infiltrating rain faster, resisting stress without constant intervention. The weather didn't suddenly become kinder. The system crossed a threshold where the past finally began multiplying the present.

That threshold is what compounding looks like in biology.

The Fibonacci sequence is the simplest, most recognizable expression of this kind of growth.

And you don't need to "be good at maths" to use it.

You only need to understand the order.

The Math of Grace

The Fibonacci sequence begins like this:

1, 1, 2, 3, 5, 8, 13, 21…

Each number is the sum of the two before it.

That's the whole rule.

It looks almost too simple to matter—until you realise it describes how a system grows without exhausting itself. Nothing is wasted. Nothing starts over. Each step carries the memory of the last two. Growth is not forced; it is built.

Now watch what happens when you divide one number by the one before it:

8 ÷ 5 = 1.6

13 ÷ 8 = 1.625

55 ÷ 34 = 1.617

89 ÷ 55 = 1.618

As the sequence grows, that ratio stabilises around **1.618**—the Golden Ratio, phi. This ratio shows up repeatedly in natural systems because it represents a structure that expands while keeping its integrity.

It grows without collapsing.

That's the point.

In soil, the same logic appears in a loop you can see if you know where to look:

Better aggregates create more pore space.
More pore space allows more fungal growth.
More fungal growth produces more biological glue.
More glue creates stronger aggregates.
Stronger aggregates hold more water, house more life, and build more stability.

This is not a metaphor. It is a mechanism.

The system builds itself by remembering itself.

Why Your Progress Feels Slow

Here is the part most farmers are never told:

In the early numbers of the Fibonacci sequence, the gains are small.

1, 1, 2, 3…

It feels like nothing is happening. This is where most people quit. They decide regenerative principles are "nice ideas" but don't scale, don't cashflow, don't hold under pressure.

But the Fibonacci sequence doesn't grow like a straight line. It bends.

When it bends, the same proportional gain produces a dramatically larger result because the base it's working on is larger.

The compounding didn't suddenly begin.

It was accumulating quietly until it became visible.

Your job is not to force the bend.

Your job is to build the base that makes bending possible.

Which brings us to measurement.

Because if you measure a compounding system with a linear ruler, you will always feel like you're failing—right up until the moment the system turns and you suddenly look "lucky."

Luck is rarely the cause.

Measurement is.

CH. 2 FIELD DIAGNOSTICS

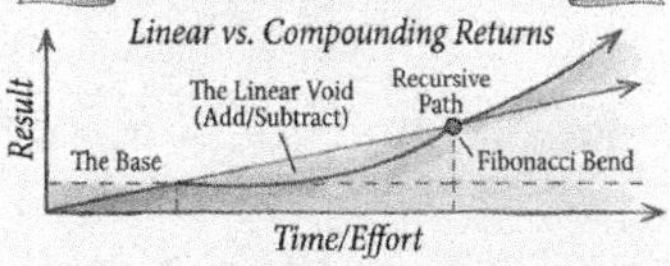

TABLE 2.1 — SOIL RESONANCE SCORE (SRS) QUICK SCORECARD

How to use: Choose a score (0–10), multiply by weight, sum totals.

1) STRUCTURE (x8)	0–2 Clod slakes fast... 6–8 Clod holds several... 3–5 Clod holds briefly... 9–10 Clod intact...	Your score: __ × 8 = _____
2) ARMOR (x5)	0–2 Bare soil common, long gaps 6–8 Consistent cover; few bare 3–5 Partial cover, seasonal gaps 9–10 365-day cover, dense shield	Your score: __ × 5 = _____
3) DIVERSITY (x3)	0–2 Monoculture / 1–2... 6–8 5+ species, multiple roles 3–5 3–4 species, some function 9–10 8+ species, full breadth	Your score: __ × 3 = _____
4) LIFE (x2)	0–2 Few worms, threads rare 6–8 Worms common, threads visible 3–5 Some worms, thin biology 9–10 Abundant soil life, rapid cycling	__ × 2 = _____
5) NUTRIENTS (x1)	0–2 System depends on balance 6–8 Mostly biological mineralisation 3–5 Partial biological + supplennent 9–10 Nutrients largely supplied...	__ × 1 = _____
	Your Soil Resonance Score (SRS): **baseline total:** _____ / 190	

Date: __________ Field/Paddock: ______________ Notes: ____________________

A New Ruler: The Soil Resonance Score (SRS)

The Linear Void measures what you extract.

Bushels per acre. Dollars per hectare. How green the crop looks from the road.

These measures aren't useless—but they are incomplete. They don't tell you whether the land is becoming more capable, or whether you're simply pushing harder.

The Soil Resonance Score measures what you are **building**.

It is a simple framework that forces your attention onto the parts of the system that determine whether soil compounds or collapses. It does this using Fibonacci weights—not to be clever, but to reflect what matters most in a living system.

Here is the formula:

SRS = (8 × Structure) + (5 × Armor) + (3 × Diversity) + (2 × Life) + (1 × Nutrients)

Notice the inversion.

The highest weight goes to what is most foundational.
The lowest weight goes to what the Linear Void obsesses over.

This is not moralising. It is mechanics.

A high nutrient program cannot rescue collapsed architecture.
A perfect spray schedule cannot substitute for pore space.
A field cannot compound without rooms.

So the SRS does what good measurement always does:

It tells you where to invest.

How to Score Without Overthinking

Score each horizon from **0 to 10**. Then multiply by its weight.

Structure (×8)

How intact is your pore space and aggregate stability?
Use your Slake Test. Use infiltration after rain.

Armor (×5)

How covered is the soil surface—living or dead—throughout the year?
If you can see bare earth for long stretches, your armor is failing.

Diversity (×3)

How many functional plant roles are present at any one time?
Not just species count—function: nitrogen fixers, deep roots, carbon builders, pollinator bridges, mycorrhizal hosts.

Life (×2)

Do you see evidence of biology—worms, fungal threads, arthropods, smell, crumb structure?

Nutrients (×1)

Are nutrients being delivered biologically, or purchased and forced? This is not a purity test. It's a diagnosis of dependency.

Now add your weighted totals. That number is your baseline.

It isn't a judgement.

It's a map.

The Inversion Point

In every compounding system, there is a threshold where effort begins to decline while results improve.

In soil, this is the moment when the system starts doing more work than you do.

In SRS terms, this often happens when your score approaches roughly **55–65%** of the maximum. Not because the number is magical, but because by that point the foundations are strong enough that the system starts expressing its own intelligence consistently:

- water infiltrates instead of running off
- soil temperature stabilises
- pest pressure reduces without constant fighting
- nutrient delivery shifts from purchased to biological
- your interventions become smaller because the land becomes more capable

A forest does not require a fertiliser budget. It produces abundance on sunlight, water, and recursion.

That is what soil wants to do when its architecture is intact.

The mistake the Linear Void encourages is beginning at the top of the stack.

Chemistry first. Control first. Force first.

But the Fibonacci logic is clear: build the base.

Structure first. Armor second.

Then diversity and life.

Chemistry last.

Not because chemistry is evil.

Because chemistry is the least powerful lever in a collapsed system.

A Farmer With the Wrong Ruler

Consider someone like **Priya**, a composite of the data driven stewards who land here after years of doing "the right practices" with the wrong measurement.

She's no till. She plants cover crops. She's reduced sprays. But she's measuring success season by season and wondering why it's still not paying.

The SRS reveals the truth:

She improved diversity and biology—but her structure and armor are still weak, and those are the highest weights.

Her gains weren't imaginary.

They were just trapped in the base of the curve.

Once she moves her effort toward architecture and protection, the curve begins to bend. The same practices finally compound. The system stops resetting.

She wasn't behind.

She was measuring the wrong thing.

Your Field Protocol: The Math in Practice

Do not let this chapter become theory.

Use it.

1) Calculate your baseline SRS today

Five scores. Multiply by weights. Add them up. Write the total and date.

2) Identify the highest leverage improvement

Look at your lowest score among the high weights:

- Structure (×8)
- Armor (×5)

Ask: *If I raised this by three points, what happens?*

A 3 point rise in Structure = **24 points** on your SRS.

A 3 point rise in Nutrients = **3 points.**

The math tells you where your next season should focus.

3) Choose one action that protects compounding

Pick one change for the next 14 days that protects the base:

- reduce one unnecessary disturbance pass
- close one bare soil window
- establish living roots sooner after harvest
- increase residue cover where soil bakes

Small change, high weight. That's the Fibonacci approach.

4) Set a 12 month target

Add **25–35 points** to your baseline SRS.

Then identify the *two horizons* that will deliver most of that change.

Those become your focus.

You are not trying to fix everything.

You are building in the correct order.

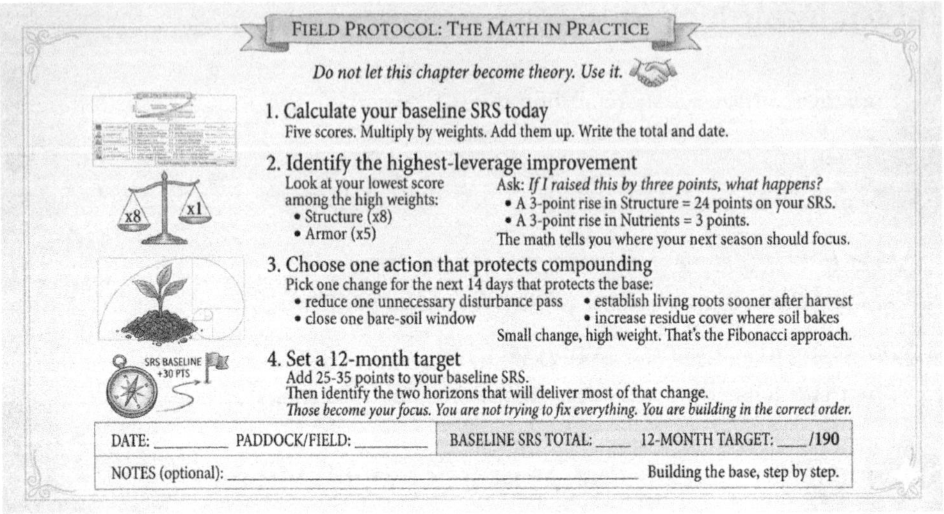

The Only Lesson You Need Right Now

The tired mind counts one and one again, and mistakes that for progress.

The living system remembers.

Your land is not asking for more effort.

It is asking for the right sequence.

Build rooms before you decorate them.

Protect structure before you chase chemistry.

The Fibonacci Pulse is already running underneath your feet.

This chapter simply gives you a ruler that matches it.

In the next chapter, we go into the night—into the **Circadian Inhale**—where the land does some of its most important work while you sleep, and where timing becomes a form of stewardship.

Chapter 3: The Circadian Inhale: Timing the Earth's Breath

The Ancestor Speaks:

"My child, the sun dips now below the far hills, painting the sky in colours we have no words for. The day's toil is done. The tools are laid to rest. And most would say, 'The field sleeps.' But come with me, out into the quiet. Take my hand. Can you feel the subtle shift in the air? That gentle settling, a cool sigh against your cheek?

"We walk the edge of the field, just as your great-grandmother taught me, under the rising moon. Feel the earth beneath your bare feet. During the long day, she laboured, exhaling moisture into the hungry sky, feeding the clouds with her bounty. It is loud, visible work, the visible rush of sun and growth. But now, as the stars begin to prick through the fading light, the true, silent work begins. Can you feel it? The earth is not sleeping, my child. She is drawing a great, slow breath. This is her Circadian Inhale.

"Place your hand flat upon this patch of soil where the green cover rests like a warm blanket. Feel the slight dampness, the cool promise. As the air above chills, it becomes heavy, sinking down, gently pushing into the myriad tiny rooms and pathways that the Great Weaver

has so diligently built. This is the Atmospheric Draw, bringing the very essence of the sky – the hidden nourishment, the breath of life – deep into the earth's heart, feeding the silent multitude who dwell there.

"Now, feel deeper. The soil below, it still holds the sun's warmth. The air above, it cools faster. This dance of warm and cool, it is the magic of the Dew-Point Alchemy. The moisture that rose to the heavens during the day, it begins to condense now, jewel-like, not just on the leaf, but it is drawn down, into the very architecture of the soil. It is an internal rain, a watering event orchestrated by the moon and the cool night air, happening without a single drop falling from the sky. It is the Earth drinking, silently, replenishing herself from the atmosphere itself. My grandmother, she would always say, 'A healthy field drinks the dew like a thirsty child at its mother's breast.' If your earth lies bare, exposed, or if its castle walls are broken, how can it drink this free, unending gift? It would simply run off or vanish with the dawn.

"And below, in the rich dark, while the dew is drawn, the roots of the plants, they are whispering to their fungal partners. This is the Carbon Anchor. The liquid sunshine, the sugars made through the day, they flow freely now into the mycelial threads. It is the nightly transaction, the giving and receiving, binding the sky's light deep into the earth's structure, strengthening the very fabric of existence. This is the Fibonacci Pulse in its deepest, most secret rhythm, compounding resilience, building the future while we sleep.

"My son, my daughter, many have forgotten this. They fill the night with the clatter of machines, the harsh glow of lights, tearing at the very moment the Earth needs to breathe her deepest. They disrupt the draw, break the alchemy, sever the anchor, blind to the harm they cause. They are blind because they do not listen to the quiet.

"Now, take a deep breath yourself. Feel the cool air fill your own lungs. The earth, too, breathes. Your task is not to control this breath, but to protect it. To allow it. The Seventh Generation will inherit the consequence of whether you let the Earth breathe freely through the night, or whether you chose to silence her most sacred work. Can you feel the profound peace that comes from simply allowing, from trusting the ancient rhythm?

"Close your eyes again. Feel the coolness on your skin, the subtle dampness. Listen for the silent intake of the Earth's breath. It is happening, right now. Can you feel it pulling you into its ancient rhythm, connecting you to every living thing?"

Meditation: The labour of night

The Circadian Inhale

"The sun-eye wanes, the earth's deep thermal sigh,
Drawing a silver breath from humid sky.
Attend the moon-tide, know the dew's embrace,
To drink from air when clouds have left no trace.
For night, the silent mother, brings forth grace."

The Circadian Inhale: Timing the Earth's Breath

The sun goes down and most farmers go inside. The tractor keys get hung by the door. The day is done. But the land is just getting started.

Everything we've built so far, the cathedral of pore space, the glomalin masonry, the Fibonacci weighting of your SRS, all of it exists to serve a process that most stewards have never witnessed because it happens while they're asleep. The soil breathes at night. Not as a metaphor. As a measurable, mechanical, biological event that draws gases, moisture, and carbon deep into the earth's architecture during the cooling hours between sunset and dawn. It's the most productive shift on your farm, and it runs without you. Your only job is to stop getting in the way of it.

This chapter is about learning to read the night.

The Planetary Pulse

During the day, the soil exhales. The sun drives the great engine of transpiration upward. Plants pull water from deep horizons and release it as vapor into the sky. The surface heats. Gases expand and rise. Moisture evaporates. The soil sweats. This daytime exhale is active, visible, and loud. It's the part of the cycle that modern agriculture has built its entire schedule around, because it's the part you can see from a tractor cab.

But a system that only exhales eventually collapses.

When the sun drops below the horizon and the surface begins to cool, the pressure gradients of the planet shift. The air near the ground becomes denser. The soil surface, warmer than the air above it, creates a gradient that pulls the cooling atmosphere downward. The earth stops pushing and starts pulling. This is the Circadian Inhale, the nightly breathing cycle of the planet's soil, and it operates through three distinct, sequential processes that are worth understanding in detail because each one can be either supported or sabotaged by what you do, or don't do, in the hours around it.

The first process is the Atmospheric Draw. As the cooling air sinks, it flows into the Silent Architecture, the pore network you've been protecting and rebuilding since Chapter 1. It carries with it nitrogen, oxygen, and trace gases that the subsoil microbial community has been waiting for all day. The bacteria responsible for nitrogen fixation are particularly active during this period. They're not waiting for you to apply a bag of nitrogen. They're waiting for the air to arrive through the pore channels you've either kept open or collapsed. A soil with an Architecture score of 7 or 8 receives this atmospheric delivery efficiently, routing gases deep into the profile where the biology lives. A soil with an Architecture score of 2 or 3 has no channels to draw through. The air sits on the surface, the microbes go hungry, and the nitrogen fixation that should be happening for free doesn't happen at all.

DIAGRAM OF THE CIRCADIAN INHALE (NIGHTTIME SHIFT)

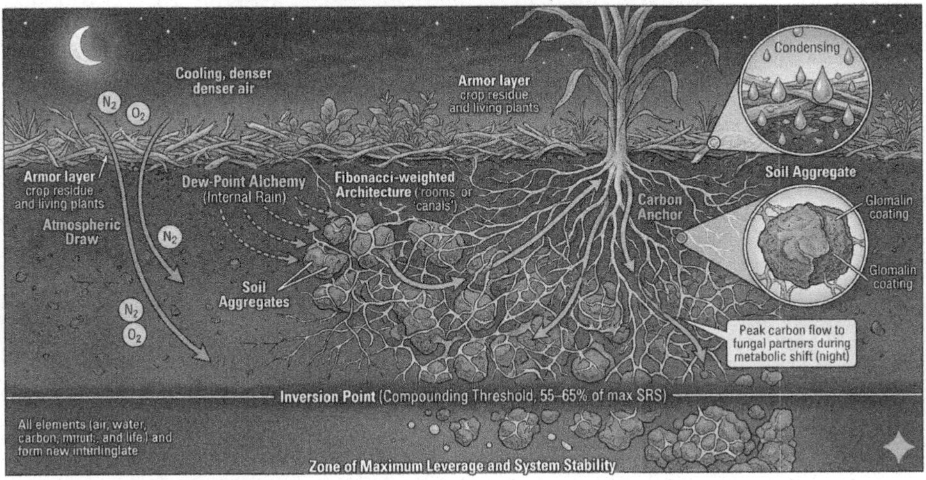

Figure 3.1. The circadian inhale: the nightly pulse of the soil.

The second process is what we'll call the Dew-Point Alchemy. Moisture that was driven upward into the atmosphere during the day's transpiration doesn't disappear. As the air cools toward its dew point, that moisture begins to condense. In a healthy, armoured soil, this condensation doesn't just form a thin layer of dew on the surface. It's actively pulled downward into the aggregate structure by capillary tension and the temperature gradient between the warm subsoil and the cooling surface. This is Internal Rain. It's a watering event that happens every single night, drought or no drought, without a pump, without an irrigation system, without any intervention from you. The only requirement is that the architecture is open enough to draw the moisture in and the armour is intact enough to prevent it from immediately evaporating back out at first light.

The third process is the Carbon Anchor. During the day, plants produce liquid carbon through photosynthesis, the sugar-rich exudates that flow down through the roots into the fungal network. But the peak delivery of this carbon to the fungal partners doesn't happen during the heat of the day. It happens at night, when the metabolic pressure in the root system shifts and the liquid carbon flows most freely into the mycorrhizal threads. The fungal network, which we'll map in detail in Chapter 4, does its most active weaving in the dark, incorporating that liquid carbon into the stable, glomalin-bound aggregate structure of the soil. This is how carbon gets anchored permanently into the earth. Not through a machine. Not through an additive. Through a nightly biological transaction between plant root and fungal thread that has been running on this same schedule for four hundred million years.

These three processes, the Atmospheric Draw, the Dew-Point Alchemy, and the Carbon Anchor, are not independent events. They're a sequence. Each one depends on the conditions created by the one before it. The architecture must be open for the draw to work. The draw must pull cool air in to create the temperature gradient that drives the dew-point condensation. The condensation must move moisture into the aggregates to create the aqueous environment where the carbon anchor operates most efficiently. Interrupt any one of these steps and the whole sequence degrades. The compounding stops. The Fibonacci Pulse skips a beat.

The Linear Void doesn't know this sequence exists. Its schedule is built entirely around the daytime exhale. It plants by the calendar, sprays by the calendar, and runs machinery by the clock on the wall. The inhale is invisible to it because the inhale happens in the dark, in silence, without any visible activity to manage or optimize. And so the Void schedules its most disruptive operations, the nighttime plowing, the late-evening chemical applications, the pre-dawn tillage runs to beat the heat, precisely during the hours when the soil's most vital work is underway. It's not malicious. It's blind. But the biological cost is the same either way.

The Dew-Point Alchemy

Water is the most underestimated resource on most farms. Not because it's scarce, though in many regions it is, but because the system that should be capturing and storing it has been compromised. Most irrigation bills aren't water problems. They're architecture problems. They're the financial consequence of a soil that can no longer pull moisture from the atmosphere at night and store it in its aggregate structure during the day.

The mechanics of dew-point condensation inside a healthy soil profile are worth slowing down to understand, because once you see them clearly, the logic of everything that protects the inhale becomes obvious.

When the soil surface cools at night, it reaches the dew point, the temperature at which the water vapor in the air above it begins to condense into liquid. In a bare, compacted soil, this condensation forms on the hard surface and either evaporates again at dawn or runs off the sealed surface without entering the profile. The soil gets wet on top and stays dry underneath. The plants don't benefit. The microbes don't benefit. The moisture event is wasted.

In a healthy, armoured soil with intact aggregate structure, the same condensation event produces a completely different result. The temperature gradient between the warm subsoil, which retains heat from the day, and the cooling surface creates a pressure differential that actively draws the condensing moisture downward into the pore network. The aggregate surfaces, coated in glomalin, are hydrophilic at the micro scale, meaning they attract and hold water molecules. As moisture enters the pore channels, capillary tension pulls it deeper. The soil is drinking the atmosphere. On a night with 80% relative humidity, a well-structured soil with a thick armor layer can absorb the equivalent of two to four millimetres of rainfall through this condensation mechanism alone. Over a growing season of 150 nights, that adds up to 300 to 600 millimetres of water captured from air that a bare, compacted neighbour field completely ignored.

That's not a trivial number. In a semi-arid environment, 300 millimetres is the difference between a crop that survives a dry spell and one that doesn't.

The armour layer is the critical variable here. A thick mat of surface residue, whether it's crimped cover crop biomass, standing stubble, or a dense living mulch, does two things simultaneously for the dew-point process. First, it insulates the soil surface, slowing the rate at which the surface temperature drops. This extends the period during which the temperature gradient between surface and subsoil is steepest, which is exactly when the capillary draw is strongest. Second, it physically captures condensing moisture on its own surface, holding it in contact with the soil long enough for it to be drawn in rather than evaporating at first light. The armour is not just protecting the architecture from mechanical and thermal damage. It's actively participating in the moisture capture process every single night.

This is why the Armor horizon carries a Fibonacci weight of 5 in your SRS. It's not just a shield. It's a collection mechanism. A field with 100% armour coverage and open aggregate structure is running a passive irrigation system every night of the year, free of charge, powered entirely by the temperature differential between the warm earth and the cooling sky. A field with bare soil is leaving that system switched off and paying for the water it loses through an irrigation pump instead.

To leverage this process practically, you need to know your soil's temperature gradient during the peak inhale period. This is specific and measurable. You'll need two inexpensive soil thermometers, the kind with a probe you push into the ground. Place one at a depth of two inches and one at a depth of eight inches. Take readings at 10 PM, at 2 AM, and at 6 AM for three consecutive nights. What you're looking for is the spread between the two readings at 2 AM. If the eight-inch reading is more than four degrees Celsius warmer than the two-inch reading, your temperature gradient is strong enough to drive meaningful capillary draw. If the spread is less than two degrees, your surface armour is inadequate and the gradient is being lost to rapid surface cooling. This tells you specifically whether your armour is thick enough to maintain the gradient, not just whether it exists.

A gradient reading of less than two degrees at 2 AM is an immediate action signal. It means your armour layer needs to be thicker before the next growing season. The target is a minimum of five centimetres of surface residue, measured as compressed depth after settling, not the standing height of the cover crop before crimping. Five centimetres of compressed brown residue maintains the gradient. Three centimetres doesn't. This isn't an estimate. It's the threshold below which the dew-point mechanism loses most of its effectiveness.

The Internal Rain effect is the Fibonacci Pulse made liquid. Every night, if the architecture is open and the armour is intact, your soil is compounding its water reserves without any input from you. That compounding, night after night, season after season, is what produces the drought resilience that looks miraculous from the outside but is simply physics from the inside. The soil isn't lucky. It's structured correctly. And structure, as we've seen throughout this book, is the highest-weighted variable in the entire system.

Stewardship by the Moon

There's an old poem embedded in the source materials for this book. It goes like this: "Don't wake the soil before the dew is dry, for night is when the deep roots touch the sky." Whoever wrote that line understood something that most modern field schedules ignore completely: the timing of your actions on the land matters as much as the actions themselves.

A seed planted during the dew-draw, when the soil is actively pulling moisture downward, has a fundamentally different start to its life than a seed planted in the dry heat of midday. During the inhale, the soil is creating the exact conditions that a germinating seed needs: moisture moving toward the root zone, cool temperatures at the surface, and gas exchange happening actively through the pore channels. The seed is pulled into the heartbeat of the soil rather than sitting on top of a system that's pushing everything upward. Germination rates improve. Root establishment is faster. The seedling's first mycorrhizal contacts happen sooner because the fungal threads are most active during the same hours the moisture is moving.

This isn't folklore. It's the logical consequence of planting in alignment with the soil's natural pressure gradients rather than against them.

Seeding during the dew-draw means timing your planting operations to finish by late afternoon so that the seed is in the ground and the surface is undisturbed before the inhale begins. For most temperate regions, the inhale's peak draw occurs between 10 PM and 4 AM. Any mechanical disturbance of the soil surface during these hours, whether from seeding equipment, tillage implements, or even heavy foot traffic across a wet surface, disrupts the capillary channels that are actively drawing moisture downward. You're not just compacting the surface.

You're physically interrupting the suction mechanism at the moment it's working hardest.

Think about what midnight plowing actually does to this system. The tractor moves across the field during the peak of the Atmospheric Draw. The implement shatters the aggregate structure that the temperature gradient is using as its channel network. The soil surface, which was maintaining a precise temperature gradient to drive capillary draw, is inverted and exposed to the night air, immediately disrupting the gradient. The fungal threads that were in the middle of their most active carbon-anchoring period are severed. The moisture that was being drawn downward is now exposed at the surface and will evaporate at first light. In a single pass, you've interrupted all three processes of the Circadian Inhale simultaneously. You've forced a chemical exhale when the biology was trying to draw its deepest breath.

Nighttime chemical applications are equally disruptive, though for different reasons. Synthetic inputs applied during the peak inhale period are drawn into the soil profile along with the atmospheric gases and moisture. This sounds efficient, but the concentration effect in the pore channels during the draw period means the chemistry reaches the microbial community at higher concentrations than it would during the day, when the soil is exhaling and diluting inputs upward. Fungicides applied at 11 PM don't just sit on the surface. They're actively delivered into the fungal network by the same capillary mechanism that should be carrying carbon and moisture. The inhale that should be feeding the biology is instead delivering the chemistry that suppresses it.

Consider a composite scenario based on the patterns described throughout this book. A farmer, call him Thomas, 44 years old, manages 200 acres of mixed cropping in a region with hot summers and moderate rainfall. Thomas has been no-till for three years. His cover crop mixes are solid. His SRS Architecture score has climbed from 2 to 5. But he's been struggling with inconsistent germination in his autumn seeding, and his soil moisture levels are lower than he expects given the rainfall records. Thomas has been doing his seeding runs starting at 4 PM and finishing around midnight, because that's when the soil is dry enough to carry the equipment without rutting. He's technically correct that the soil is drier at that time. But he's finishing his operations precisely when the inhale is beginning, leaving the freshly disturbed seedbed exposed during the peak moisture-draw period. The seeds are sitting in disrupted channels during the hours when those channels should be pulling moisture toward them. His germination problem isn't a seed problem. It's a timing problem. Shifting his seeding window to finish by 3 PM gives the disturbed surface four to five hours to begin settling before the inhale starts, and his autumn germination rates improve measurably in the following season. Same seed. Same soil. Different timing. Different result.

Respecting the inhale also means rethinking what "working the land" looks like during the nighttime hours. The most productive thing you can do between 10 PM and 4 AM is nothing. Walk the field if you want to. Observe. Put your hand on a patch of armoured soil and feel the difference between it and a patch of bare ground nearby. The armoured soil will feel slightly warmer, slightly damp, slightly alive in a way that's hard to describe but immediately recognizable once you've felt it. That sensation is the Dew-Point Alchemy in progress. That's the Internal Rain being drawn in. That's your SRS Architecture score compounding quietly in the dark, without any help from you, because you had the wisdom to leave it alone.

The Resonant Steward doesn't fill the night with noise and diesel. They use the quiet hours to observe the Soil Resonance Score in its most honest state, when the system is running freely, without interruption, on its own source code. Does the surface feel tight and sealed, or does it feel like something breathing? Does the armoured patch feel alive while the bare patch feels cold and dead? These aren't poetic impressions. They're diagnostic signals. They tell you whether the inhale is happening, whether the architecture is open, whether the armour is thick enough, whether the Fibonacci Pulse is beating or skipping.

In the Linear Void, time is a commodity to be spent efficiently. Every hour of darkness is an hour you're not producing. The night shift exists to extend the day. But in the Spiral, time is a frequency. The night has its own work to do, and that work is more valuable than anything a tractor can accomplish during those hours. Your job after dark is to be a protector of the breath, not an interrupter of it.

Putting It Into Practice

The Circadian Inhale isn't a concept to admire from a distance. It's a system to align with. And alignment requires specific, observable changes to how you time your field operations, how you measure your soil's nighttime behaviour, and how you protect the conditions that allow the inhale to do its work.

Here are six concrete actions. Each one has a specific, measurable outcome tied directly to your SRS and to the three processes of the inhale.

The first action is to take your soil temperature gradient readings tonight. Push a probe thermometer to two inches and another to eight inches in a representative location on your land. Record the readings at 10 PM, 2 AM, and 6 AM. Do this for three consecutive nights. Calculate the average spread between the two-inch and eight-inch readings at 2 AM. If the spread is less than two degrees Celsius, your armor layer is insufficient to maintain the gradient that drives the dew-point draw. Write down the number. This is your Gradient Score, and it tells you more about your nighttime moisture capture than any daytime soil test can.

The second action is to map your current field operation schedule against the inhale window. Write down every operation you've performed in the last month and the time of day you performed it. Mark any operation that occurred between 9 PM and 5 AM with a red flag. For each flagged operation, write down what it was trying to achieve and whether the same result could have been achieved by finishing the operation before 4 PM instead. This exercise often reveals that nighttime operations are scheduled for convenience or habit, not necessity. Identify which ones can be shifted to daytime windows without compromising the outcome.

The third action is to measure your actual armor depth right now. Go to your most recently managed field and press a ruler straight down into the surface residue layer until it contacts the soil. Read the compressed depth. If it's less than five centimeters, your dew-point mechanism is operating below its effective threshold. Calculate how much additional biomass you'd need to add to reach five centimeters. A single pass of a high-biomass cover crop like cereal rye or hairy vetch, crimped and left flat, typically adds three to six centimeters of compressed residue. If you're below threshold, this is your armor prescription for the next planting window.

The fourth action is to redesign your next seeding operation around the inhale window. Plan to have all seeding equipment in the field by 8 AM and all operations complete by 3 PM. This gives the disturbed seedbed a minimum of six to seven hours of settling time before the inhale begins. If your field conditions don't allow morning seeding due to moisture, identify whether a light surface rolling the evening before would firm the seedbed enough to carry equipment without rutting in the morning. The goal is to have the seed in the ground, the surface settled, and the equipment off the field before the atmospheric draw begins.

The fifth action is to identify any synthetic inputs currently scheduled for evening application and reschedule them to a morning window between 7 AM and 11 AM. During morning hours, the soil is transitioning from inhale to exhale. The pore channels are still open from the night's draw but the capillary pressure is reversing, meaning inputs applied at this time are less likely to be concentrated into the deep profile by the inhale mechanism. If a product requires application during a specific temperature window that only exists at night, consult with your agronomist about whether a lower-concentration, biologically compatible alternative could achieve the same result without disrupting the inhale.

The sixth action is to walk your land at 2 AM on one night in the next two weeks. Not to work. Not to check on anything specific. Go out with a thermometer, a small flashlight, and your hands. Find a patch of armored soil and a patch of bare soil within twenty meters of each other. Press your palm flat on each surface for thirty seconds. Feel the temperature difference. Feel whether one surface seems to be releasing vapor while the other seems dry and inert. Crouch down and hold your face close to the armored surface. On a night with good humidity, you may be able to feel the faint coolness of condensing moisture being drawn in. This single experience, hands on the earth at 2 AM, will do more to shift your understanding of the Circadian Inhale than any page of this book. The inhale is real. It's physical. It's happening right now on every acre of well-managed land in the world. You just need to be awake to witness it once.

The cathedral is built. The rooms are open. The mason is at work in the dark. Your SRS Architecture score is the foundation, your Armor score is the skin that holds the gradient, and the Circadian Inhale is the breath that moves through both of them every night, anchoring carbon, drawing moisture, and feeding the biological network that the next chapter will map in full. The soil's most vital work happens while the world is asleep. The Resonant Steward's most important decision is simply to let it.

Chapter 4: The Mycelial Internet: Subterranean Intelligence

The Ancestor Speaks:

"Come here, my child. Not to the sun-drenched surface, but to the cool, quiet dark. Kneel beside me, here where the oldest trees murmur secrets to the earth. Place your hand gently on the soil. Close your eyes. And listen. Not for sound, but for whispers. Can you feel the faint tremor, the soft, ceaseless exchange beneath your palm?

"They told us, in the time of forgetting, that each plant stood alone, a solitary warrior battling for a scrap of light, a drop of water. They told us the earth was a battleground, a place of endless competition. It was the harsh story of the ***Linear Void****, where every living thing was isolated, a disconnected island. And in their ignorance, they drove their steel teeth into the earth, severing, silencing, and leaving the plants to wither in their loneliness. Can you feel the deep sorrow of that isolation, the aching silence where once there was a song?*

"But I remember, and my ancestors before me remembered, that the earth is not a collection of isolated lives. It is a vast, living tapestry, woven by the ***Great Weaver****. Beneath us, stretching further than any road you will ever walk, thinner than the finest hair from your head,*

is a network of silver threads. It is the very mind of the meadow, the nervous system of the forest. It connects the tiny seedling to the ancient oak, sharing strength, sharing wisdom, sending urgent messages from one end of the landscape to the other. It is the earth's own ***Mycelial Internet****, built not with copper and light, but with living breath and ancient purpose.*

"When a plant stands in need, the Weaver senses it. When a hidden trove of water or mineral lies too deep for a root to reach, the Weaver knows. And the threads, always growing, always seeking, will reach, will find, will gather, carrying the hidden gifts of the earth directly to the thirsty plant. In return, the plant, bathed in the sun's light, offers its own sweet liquid carbon – the concentrated sunshine – to the Weaver. It is a sacred trade, an ancient pact of reciprocity, the beating heart of the ***Carbon Exhale****. Each giving, each receiving, each growing stronger, not just for themselves, but for the whole. This is the* ***Fibonacci Pulse*** *in its most intimate dance, compounding life, thread by silver thread.*

"Our people never broke these threads without deep reverence, without apology. For to cut the Weaver was to silence the earth's whispers, to make the plants blind and deaf, to condemn them to a solitude they were never meant to endure. The new ways, the churning steel, they tore this network to shreds, leaving behind not a connected intelligence, but a fragmented memory. They created a profound silence where there should have been a symphony.

"Now, child, open your eyes. Look at the living green around you. Then close them again. Can you almost see the shimmering network, feeling its presence, its patient resilience? Can you imagine the sheer intelligence flowing through those threads, the silent conversation of a million roots? The ***Seventh Generation*** *watches, not just for the harvest you reap, but for the connections you honour.*

Will you listen to the Weaver's song, or will you allow the silence to deepen? The choice to mend these broken ties, to allow the network to heal, lies in your gentle hand. Feel the power of that connection, flowing through you, from the deepest earth to the highest sky."

Prelude: Threads that listen

The Mycelial Internet (Part I)

" The nerve-web's white, the meadow's waking mind,
Where liquid light for mineral depth is signed.
To cleave the earth is to sever every tie,
And leave the roots to starve beneath a silent sky.
Protect the Weaver; the Weaver feeds all life."

The Mycelial Internet: Subterranean Intelligence

The cathedral needs a nervous system

By now you've seen the architecture. You've learned that soil is not a pile of minerals—it is a building made of space. The pores are the rooms. The aggregates are the doorways. The armor keeps the rooms from collapsing.

But a cathedral without communication is still vulnerable. One wall cracks and the rest of the building doesn't "know." There is structure, but no intelligence.

Soil has intelligence.

Not in the sentimental sense, but in the literal, functional sense: a living system that senses, responds, distributes resources, and builds resilience through connection.

That intelligence lives in a network older than your farm, older than your crops, older than human agriculture itself. It is made of threads finer than a human hair, and it links the roots of plants into a coordinated, cooperative system.

This is the **Mycelial Internet**.

Most stewards have farmed on top of it their entire lives without being introduced to it.

That changes now.

The end of isolation

The story we've been told about plants is a story of competition. Every stalk fighting for light. Every root racing for moisture. A field as a battlefield of individuals.

That framing suits the Linear Void because it justifies force: if everything is competing, then management becomes control—eliminate threats, choose winners, and push productivity.

But under your feet, the evidence points to a different reality.

A large proportion of plants form partnerships with **arbuscular mycorrhizal fungi**—AMF—an underground web of hyphae that extends far beyond what roots can reach alone. The fungal threads link plants to one another and to the mineral and moisture banks of the soil profile. They move resources. They carry signals. They help the system behave as one organism rather than a field of isolated bodies.

In healthy, undisturbed soil, a small amount of earth can contain an astonishing length of fungal threads. Scaled across hectares, the network becomes vast—an underground infrastructure that makes "isolation" an expensive condition for plants to endure.

Isolation has a cost.

When a plant must operate alone, it can only access the narrow boundary of soil touching its roots. That boundary depletes quickly. The plant then spends carbon building more roots just to keep feeding itself—energy diverted away from growth, defense, and resilience.

In a connected system, the plant outsources part of that work. The fungi extend its reach into pores, crevices, and mineral surfaces no root tip can enter. The plant's effective foraging zone expands dramatically. The cost of isolation falls. The return on sunlight rises.

This is the first major shift of this chapter:

You are not managing individual plants. You are managing connectivity.

THE MYCELIAL INTERNET (UNDERGROUND CONNECTIVITY)

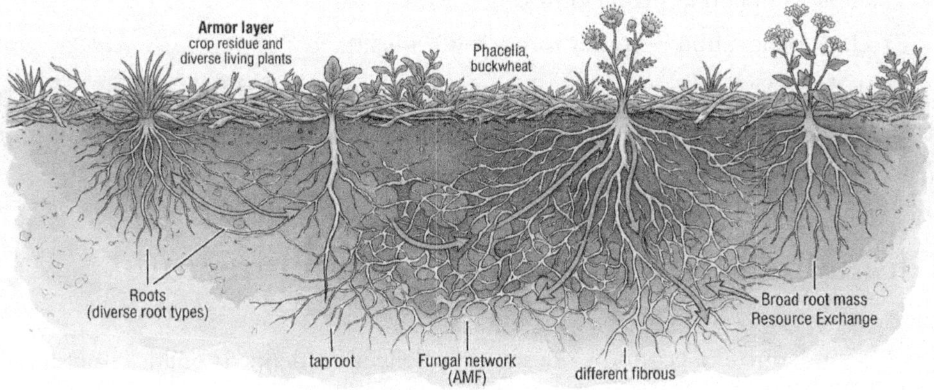

Resource and signal exchange occur through this vast, symbiotic fungal web, linking all living roots into a coordinated intelligence layer.

Figure 4.1 (Mycelial Internet)

The underground trade: liquid carbon for mineral depth

A plant is a solar factory. Through photosynthesis it turns sunlight and air into sugars—liquid carbon. In a living system, the plant releases a significant portion of that liquid carbon through its roots. This is not waste. It is investment.

Root exudates are currency.

They attract microbes. They feed fungi. They build relationships.

The Mycelial Internet responds to that currency with mineral wealth: phosphorus, zinc, magnesium, iron—nutrients held in places roots cannot reach efficiently. Fungal threads dissolve, transport, and deliver minerals to plant roots in exchange for carbon.

This exchange is the foundation of soil fertility that cannot be purchased in a bag.

And here is where the Fibonacci idea belongs—without symbols:

Living systems build tomorrow from what they built yesterday.

When the trade is active, each season leaves behind infrastructure that makes the next season easier: more fungal threads, more glomalin, more stable structure, more pore space, more infiltration, more resilience.

That is recursion in plain clothes.

Today's connection becomes tomorrow's advantage.

This is why a mature pasture, hedgerow, or undisturbed forest edge behaves like a different universe than a repeatedly disturbed field. The difference is not luck. It is continuity.

Glomalin: when trade becomes architecture

The underground economy doesn't only deliver nutrients. It builds structure.

As fungal threads grow through soil, they produce and deposit **glomalin**—a tough, carbon rich "biological glue" that binds mineral particles into stable aggregates. Those aggregates hold pore space open. Pore space enables infiltration. Infiltration enables life.

So every trade is also construction.

When plants feed fungi with carbon, fungi don't just "help" the plants. They also leave behind the mortar that keeps the cathedral standing.

This is why the Mycelial Internet is not an optional feature. It is the building crew and the communications system at the same time.

Information warfare: what tillage really destroys

Every spring, across vast acres, the same act repeats so routinely it no longer looks like destruction.

Steel enters soil. Soil is turned. The surface is made to look "ready."

But from the point of view of the underground network, that act is closer to cutting every cable in a city at once.

The buildings remain. The streets remain.

The network is silent.

A fungal thread does not stretch when it is cut. It breaks. And when a network is severed repeatedly, it can never mature. The system remains in a state of perpetual recovery—always rebuilding, never compounding.

This creates a predictable loop:

- disruption severs the network
- plants lose mineral delivery and signal sharing
- stress increases (nutrient deficiency, pests, drought sensitivity)
- the manager compensates with inputs
- inputs reduce the plant's incentive to invest in fungi
- the network rebuilds slowly
- disruption arrives again and resets it

That is the Linear Void made biological.

And the cost shows up as:

- higher nutrient bills
- higher spray dependence
- weaker resilience
- and an exhausted steward.

The two week rule: continuity is not negotiable

There is one constraint the underground system will not forgive:

no living roots.

AMF fungi cannot survive indefinitely without a host plant. When land is left bare—especially in warm conditions—the network loses its food source. The threads thin. The relationship weakens. The architecture begins to degrade.

The practical implication is simple and specific:

Do not leave a field without living roots for more than two weeks.

That two week window is not ideology. It is biology. It is the difference between a network that bridges seasons and a network that collapses into dormancy and must be rebuilt from scratch.

If you want soil to compound, you must stop resetting it.

Continuity is the invisible lever behind most "miraculous" regenerative outcomes.

A practical transition: reducing disturbance without losing control

The move away from tillage does not require perfection. It requires sequence.

If you attempt to change everything at once, the system can revolt—weed pressure, residue management, equipment mismatch. The transition must protect the network while you build new skills.

A practical phased approach:

1. **Start where weed pressure is manageable.**
   Choose one or two fields where you can realistically reduce disturbance without chaos.
2. **Terminate cover crops without tearing roots out.**
   Roll, crimp, or use the lightest possible termination method that leaves the root system in place.
3. **Keep living roots running through the gaps.**
   If the cash crop ends, a cover crop begins—fast. The goal is to keep the underground economy fed.
4. **Treat depth as sacred.**
   If you must cultivate strategically, keep disturbance shallow. Deep disturbance is deep network loss.

The goal isn't a "no till badge."
The goal is uninterrupted infrastructure.

Your next move: four actions in fourteen days

This chapter only matters if it changes what you do next.

Here are four actions with direct, observable outcomes:

1) Do a fungal thread check

In three locations—(a) continuously covered, (b) recently disturbed, (c) long bare—break apart ten aggregates over a dark surface and look for fine white threads. Record how many aggregates show visible thread.

You are measuring network presence, not theory.

2) Identify your biggest "rootless gap"

Look at your calendar. Where is your land bare and rootless? Count the days. Any gap beyond two weeks is a network die off event in slow motion.

Circle your biggest gap. That is your first place to change the system.

3) Add one strong mycorrhizal host to your next mix

If you plant cover crops, ensure at least one species in your mix is a strong mycorrhizal partner. Your goal is not green cover alone—it is carbon delivery to the network.

4) Choose one field as a three year "network sanctuary"

Commit in writing: no deep steel disturbance in the topsoil for three years. Use it as your living laboratory. Retest slake stability and fungal presence season by season.

By the end of year three, that field will teach you more than arguments ever could.

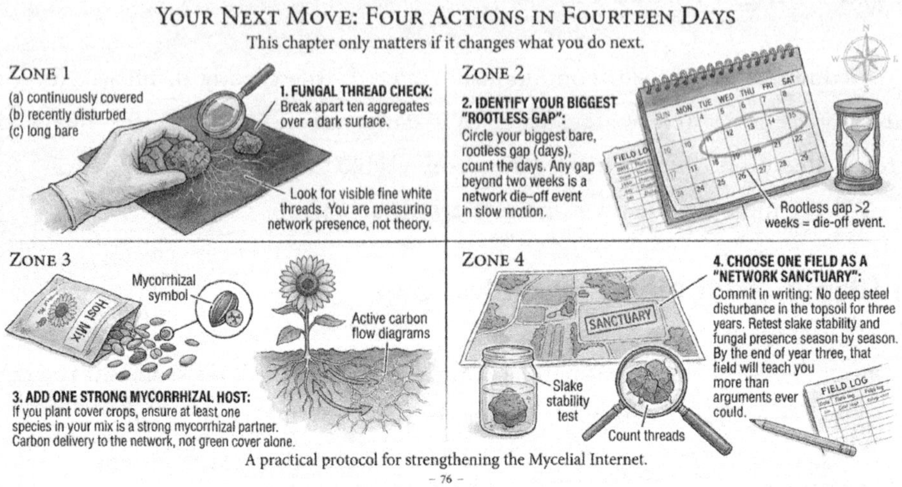

Figure 4.2 Your next four actions in fourteen days

The quiet truth of this chapter

The Mycelial Internet is not a metaphor. It is infrastructure.

It is what allows soil to remember.

It is the difference between a system that compounds and a system that resets.

And it asks only one thing of you:

Stop cutting the threads.

In the next chapter, we move to the surface—the skin of the soil—and learn why protection is not aesthetics but survival, and why Armor carries a weight that can make or break the pulse.

Chapter 5: The Living Armor: Protecting the Pulse

The Ancestor Speaks:

"My child, come stand beside me. The sun is high, a relentless eye in the sky. Look upon this field – a testament to human order, they would say. Bare. Clean. Every clod broken, every living thing swept away. They call this 'good farming.' But can you feel the heat rising from it, scorching the very air? Can you hear the Earth's silent scream? My own heart aches with it. This, my child, is not order. This is a wound. A terrible, burning wound, like a ***third-degree burn*** *upon the body of our Mother Earth.*

"Our grandfathers, in their pride, showed us these 'clean' fields. They saw prosperity. But your great-grandmother, she would squint her eyes at the shimmering heat and whisper, 'The fever is upon her. The skin is gone.' She knew that without its skin, the Earth cannot live, cannot breathe, cannot protect the precious life within. The ***Linear Void*** *taught us to strip the land bare, to expose her to the harshness, calling it 'control.' But it was only ever an act of violence, born of not seeing, not feeling.*

"Do you remember the tiny castles, the ***Silent Architecture*** *that the* ***Great Weaver*** *builds with glomalin, that precious mortar? Under this fierce sun, with no covering, that mortar softens, it breaks. It melts away like snow on a summer day. The rooms collapse. The breathing*

channels, the sacred pathways for the ***Circadian Inhale*** *you've come to understand, they seal shut. And when the rain finally comes, it does not bring solace. It brings more pain, each drop a tiny hammer shattering the last remnants of structure, washing the very lifeblood of the soil away.*

"But look now, at this other place. Here, where the brown stalks lie gently, woven with new, tender green shoots. This, my child, is the Earth's true skin. This is the ***Living Armor****. See how the sun's fierce gaze is softened, absorbed by the* ***Brown Shield*** *of the fallen plants. Feel the cool dampness beneath. Here, the life still hums. The* ***Great Weaver*** *continues its patient work, its glomalin protected, its network safe from the burning eye. The water, when it falls, is welcomed, drawn in, not fought away. The breath of the night, the precious moisture, it is held, cherished, allowed to penetrate deep into the waiting soil.*

"My people understood that the Earth, like all living things, needs its cover. A ***Green Skin*** *of living plants, drinking the sun, breathing out the life, and a* ***Brown Shield*** *of resting plants, a gentle blanket guarding against the extremes. Not one or the other, but both, in an unending dance of protection and renewal. For without this skin, the* ***Fibonacci Pulse****, the very rhythm of compounding life, cannot beat freely. It falters. It struggles. It withdraws.*

"They might call this 'messy.' They might say you are 'letting the field go.' But listen to the Earth, my child. She does not crave neatness. She craves protection. She craves the shade, the insulation, the gentle touch of a thousand leaves and stems, living and resting, always shielding her heart.

"Now, close your eyes. Feel the sun upon your own skin, then imagine the cool relief of shade. Can you almost feel the Earth's gratitude when she is covered, protected? The ***Seventh Generation*** *will ask, 'Did you cover the Earth? Did you protect her skin, so she could breathe, and live, and give?' Feel the profound peace that comes with becoming her shield, her protector. The pulse beats stronger when the armour is whole."*

Reflection: Protection as generosity

The Living Armor (Part I)

" The barren earth, a wound that screams to light,
A fevered skin that spills the rain's pure might.
Though plant life cycles, never strip the ground,
For armoured layers bring deep peace profound.
Guard the Earth's pulse; its pulse will guard your own."

The Living Armor: Protecting the Pulse

The Fever of the Bare Earth

There's a particular kind of pride that runs deep in farming culture. The pride of the clean field. You've seen it, maybe felt it yourself. That stretch of dark, bare earth after a pass of the disc. Smooth. Ordered. Every weed gone. Every clod broken down. From the road, it looks like control. It looks like competence. It looks like a farmer who has their land in hand.

The soil sees it differently.

To the biology living in the top six inches of your land, bare ground isn't a clean canvas. It's a catastrophe. When the sun strikes naked soil in summer, the surface temperature can climb past 60 degrees Celsius. That's not warm. That's hot enough to denature glomalin, the biological glue that your AMF network has been building since Chapter 1. The mortar of your cathedral doesn't just soften at that temperature. It breaks down. The rooms you've been rebuilding start to collapse again. And the microbes living in the top layer of the soil, the ones responsible for nitrogen cycling, carbon anchoring, and aggregate formation, begin to desiccate and die.

This is what we mean when we say bare soil is a third-degree burn on the body of the earth.

A burn doesn't just hurt the surface. It destroys the layers underneath. A third-degree burn removes the skin, kills the tissue beneath it, and leaves the body completely open to infection, dehydration, and temperature shock. That's exactly what happens to a bare field. The skin is gone. The architecture underneath is exposed to forces it was never designed to handle without protection. And every rain event that follows isn't a recovery. It's another injury. A raindrop hitting bare soil at terminal velocity hits with enough force to shatter a soil aggregate. Researchers have measured the impact of a single raindrop on unprotected soil as equivalent to a small explosion at the microscopic scale. The aggregate breaks apart. The fine particles seal the surface. The water can't enter. It runs off, carrying your topsoil with it.

The Linear Void trained you to see that bare, dark soil as a sign of good management. It's actually a sign of a system in fever.

Think about what happens during the Circadian Inhale we covered in Chapter 3. That nightly moisture-capture process depends entirely on a temperature gradient between the warm subsoil and the cooler surface. It depends on the pore channels being open to draw atmospheric gases and condensing moisture downward. It depends on the armor layer being thick enough to slow the rate of surface cooling so the gradient stays steep through the peak draw hours. Take the armor away and the gradient collapses. The surface cools too fast. The capillary draw weakens. The Internal Rain stops. The free irrigation system you've been building through better architecture switches itself off because the skin that regulates it has been removed.

The "clean row" obsession doesn't just cost you glomalin. It costs you water. It costs you nitrogen. It costs you the nightly biological transactions that are the foundation of your Soil Resonance Score. And it costs you all of these things simultaneously, every single day that the ground is bare, while the bill keeps running and the system keeps degrading.

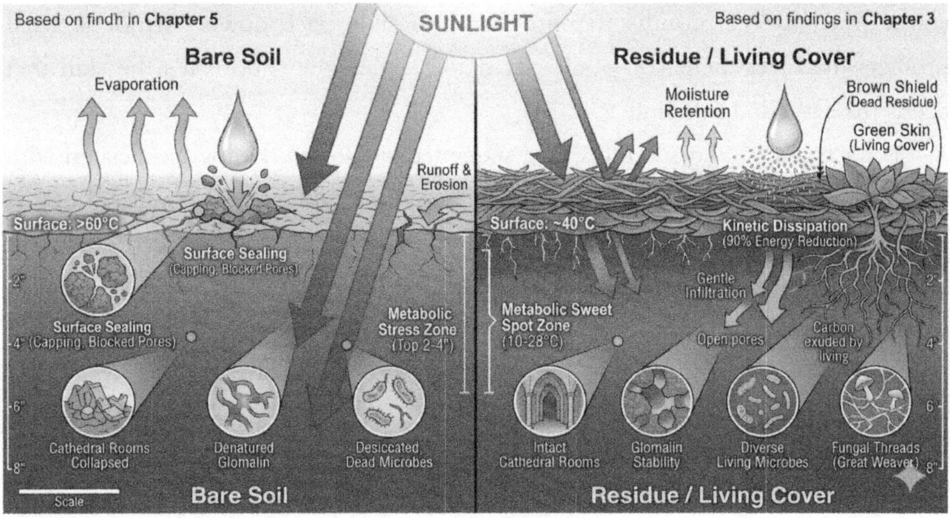

Figure 5.1 — Living and Dead Armor

Here's the shift that changes everything: stop seeing bare soil as neutral. It's not neutral. It's actively losing ground. Every day without cover is a day your SRS is moving in the wrong direction. The moment you start treating bare soil as an emergency rather than a preference, your entire approach to field management changes.

The goal, stated plainly, is this: if you can see the dirt, you are losing the pulse.

The Fibonacci Weight of Five

In your Soil Resonance Score, Armor carries a Fibonacci weight of 5. That's the second highest weight in the entire formula, sitting directly below Architecture's weight of 8. This isn't arbitrary. The weighting reflects something real about how the system compounds.

Architecture is the foundation. Without pore space, nothing else works. But Architecture without Armor is a cathedral left open to the weather. The rooms are there. The masonry is intact. But without a roof, every storm degrades the structure. Every dry spell desiccates the biology inside it. The glomalin breaks down. The aggregates weaken. The pore channels narrow. The Architecture score you worked twelve months to raise starts to slide back down. Armor is what protects the investment you've already made in the foundation. It's the skin that keeps the cathedral standing.

The weight of 5 in the Fibonacci sequence is also the number associated with the bloom, the shield, the protective layer. Look at a wild rose. Five petals. Look at a starfish. Five arms. Look at the cross-section of an apple. Five seed chambers arranged in a perfect pentagon. Five is the number nature uses when it builds something designed to protect a living center. It's the geometry of the shield. And in your SRS, it functions exactly that way.

Armor performs three distinct functions in the soil system, and each one directly supports a different component of your SRS.

The first function is thermal regulation. A thick layer of surface cover acts like a thermos. In summer, it keeps the soil cool by blocking direct solar radiation. In winter, it keeps the soil warm by insulating against frost. Both extremes matter. The biology in your soil, the bacteria, the fungi, the earthworms, the beetles, all operate within what we can call a Metabolic Sweet Spot. For most temperate soils, that's roughly between 10 and 28 degrees Celsius at a depth of two to four inches. Below that range, microbial activity slows to almost nothing. Above it, the biology starts to die. Armor is the mechanism that keeps the soil inside that window. Research on surface temperature differences between armored and bare soils consistently shows a spread of 15 to 20 degrees Celsius at the surface on a hot summer day. A bare soil hitting 62 degrees at the surface has a biology that's in crisis. An armored soil sitting at 42 degrees on the same day has a biology that's still functional. That 20-degree difference is the difference between a living system and a dying one.

The second function is kinetic dissipation. Every raindrop that hits your field is a small projectile. Without armor, that projectile hits the aggregate structure directly, shattering it. The fine clay and silt particles released by the impact then wash into the surface pores, sealing them shut. This is called surface sealing or capping, and it's one of the primary reasons water infiltration fails after rain on bare soils. The water doesn't run off because the soil is full. It runs off because the entrance has been sealed by the impact of the rain itself. Armor absorbs that kinetic energy before it reaches the soil. A five-centimeter layer of surface residue reduces raindrop impact energy at the soil surface by more than 90%. The water still arrives. But it arrives gently, flowing through the residue layer, and enters the pore network without shattering anything. The architecture stays intact. The pore channels stay open. The infiltration happens the way it's supposed to.

The third function is the fungal bridge. Dead plant material on the soil surface, the stalks, the leaves, the straw, the crimped cover crop biomass, is the slow carbon that feeds the Great Weaver through the winter months when living roots aren't providing liquid carbon through exudation. As we established in Chapter 4, AMF threads die without a carbon source. During the coldest months, when your cash crop is harvested and your cover crop hasn't yet reached peak biomass, the surface residue is the food that keeps the fungal network alive. It's the biological equivalent of a slow-release battery. It doesn't provide the high-energy liquid carbon that root exudates deliver. But it provides enough to maintain the network's thread density through the gap, so that when spring arrives and the living roots start pumping carbon again, the network is still there to receive it. Without surface residue, the network thins through winter. With it, the network holds. The Fibonacci Pulse doesn't skip a season.

A weight of 5 in your SRS means that every point you raise your Armor score produces five points of improvement in your total SRS. If your Armor score is currently a 4 and you can raise it to an 8, that's a 20-point improvement in your total score from a single horizon. No other change you can make to your management, short of dramatically improving your Architecture score, will move your SRS as efficiently as getting your soil covered and keeping it covered.

Green Skin vs. Brown Shield

Armor isn't a single thing. It's a two-layer system, and both layers serve functions that the other can't replicate. Understanding the difference between them is what separates a steward who has cover from a steward who has protection.

The first layer is what we'll call the Green Skin. This is the living armor: the cover crop that's actively growing, the perennial understory that never goes dormant, the living mulch between the rows of your cash crop. Green Skin is active armor. It shades the ground, yes. But it also does something that dead residue can't do. It pumps liquid carbon downward through its roots, feeding the fungal network in real time. It fixes nitrogen if it's a legume. It opens compacted layers with its root tips if it's a deep-rooted species. It supports pollinators and beneficial insects above ground while simultaneously feeding the biology below ground. Green Skin is the armor that pays you back while it protects you.

The second layer is the Brown Shield. This is the dead residue: the crimped cover crop mat, the standing stubble from last season's cash crop, the decomposing leaf litter from the hedgerow. Brown Shield is ancestral armor. It's the physical barrier between the sun and the soil, between the raindrop and the aggregate, between the cold air and the biology. It doesn't grow. It doesn't fix nitrogen. But it does something the Green Skin can't fully replicate: it provides continuous, unconditional protection regardless of temperature, season, or growth conditions. In the middle of a dry summer when your cover crop has gone dormant, the Brown Shield is still there. In the depths of winter when nothing is growing, the Brown Shield is still insulating the soil and feeding the fungal network its slow carbon. It's the armor that never takes a day off.

You need both. Not one or the other.

The mistake most transitioning stewards make is treating these two layers as alternatives. They plant a cover crop and assume the armor problem is solved. But a cover crop that's standing upright, with bare ground visible between the stems, is providing shade but not kinetic protection. The raindrop still hits the soil between the plants. The surface still seals. The infiltration still fails. Green Skin provides full protection only when its canopy is dense enough to intercept every raindrop before it reaches the soil. Until that canopy closes, the Brown Shield underneath it is doing the heavy lifting.

The practical target is this: at no point in the calendar year should you be able to crouch down at soil level and see a patch of bare earth larger than your hand within your managed fields. That's the standard. Not 80% cover. Not "mostly covered." One hundred percent. If you can see dirt, you have a gap in your armor. And every gap is a point where the thermal protection fails, where the kinetic impact lands unabsorbed, where the glomalin is being degraded and the surface is being sealed.

Managing the two layers in sequence is the practical approach. In the period immediately after cash crop harvest, before your cover crop is established, the Brown Shield from the previous crop's residue is your only protection. This is why residue management decisions at harvest matter so much. Chopping and spreading the residue evenly across the full width of the header is more valuable for your SRS than baling and removing it. Every tonne of residue you remove from the field is armor you're selling. The Brown Shield you leave behind is feeding the fungal network through winter, regulating the soil temperature during the frost cycles, and protecting the aggregate structure from the impact of autumn rains. Its value to your SRS is five times the weight of any nutrient it contains.

Consider a hypothetical scenario that illustrates this well. Take a composite steward we'll call Reuben, a 48-year-old mixed farmer managing 180 acres in a region with hot summers and cold winters. Reuben transitioned to no-till three years ago. His cover crop mixes are diverse and well-chosen. But every summer, he watches his soil crack in the heat and his irrigation costs stay stubbornly high despite his improving SRS Architecture score. The problem, when he finally investigates it, is a timing gap. His cover crops are established in autumn and growing well through winter. But in the gap between summer cash crop harvest in late February and cover crop germination in April, his fields are bare for six to eight weeks during the hottest part of the year. That six-week gap is enough to denature significant glomalin, kill the surface biology, and set his Architecture score back by two to three points every single summer. The solution isn't a different cover crop species. It's a Brown Shield strategy: leaving a heavier stubble height at harvest, running a light chopper to spread it evenly, and using that residue mat to bridge the gap until the autumn cover crop establishes. The Green Skin handles winter. The Brown Shield handles summer. Together, they produce the 365-day protection that his SRS Armor score needs to reach 8 instead of staying at 5.

The aesthetic shift required here is real and worth naming directly. A field covered in a thick mat of brown residue with a young cover crop pushing through it doesn't look like the clean, ordered fields that won prizes at agricultural shows for the last century. It looks messy. It looks like a farmer who hasn't cleaned up after themselves. It looks, to eyes trained by the Linear Void, like a lack of control.

What it actually looks like is a living system that's being protected. The mess is the armor. The tangle is the shield. The "dirty" look of a residue-covered field is the visual signature of a Soil Resonance Score that's climbing toward the Inversion Point. Learn to love that look. It's the most expensive thing on your farm, and it cost you nothing to produce.

Trample, Don't Tumble

Knowing you need armor is one thing. Putting it in place without destroying the architecture underneath it is another. This is where most stewards hit a practical wall. The cover crop is standing. The cash crop needs to go in. The conventional answer is to till the cover crop into the soil, incorporating the biomass and clearing the surface for planting. It feels logical. You're adding organic matter. You're making room for the new crop. You're doing what farmers have done for generations.

But you already know what tillage does to the Mycelial Internet. You know what it does to the aggregate structure. Incorporating a cover crop by tillage is the biological equivalent of burning down your house for warmth. You get a brief benefit from the released nutrients. You lose everything the living system was doing for your SRS in the process. The fungal threads are severed. The glomalin is disrupted. The pore channels collapse. And the biomass that should have been your Brown Shield for the next three months is now buried underground, where it will decompose anaerobically and produce compounds that can actually suppress germination rather than support it.

The alternative is crimping, and it's one of the most powerful single techniques available to a steward who wants to build armor without sacrificing architecture.

Crimping means rolling the standing cover crop flat using a heavy roller with a crimping pattern on its surface. The crimper doesn't cut the plant. It crushes the stem at regular intervals, severing the vascular tissue so the plant can no longer transport water or nutrients. The plant dies in place. It falls flat, creating a thick, woven mat of interlocked stems and leaves that lies directly on the soil surface. The root system stays intact underground. The fungal connections stay intact. The glomalin bonds stay intact. The pore channels stay open. And the mat of dead biomass on the surface immediately begins functioning as a Brown Shield, intercepting rainfall, regulating temperature, and feeding the fungal network as it slowly decomposes from below.

This mat is what the source materials for this book call a biological bandage. It covers the wound of the harvest gap. It shields the soil from the sun's direct radiation. It keeps the moisture in. It keeps the temperature stable. And it does all of this while the cash crop is planted directly through it, using a no-till seeder that cuts a narrow slot through the mat, places the seed at the correct depth, and closes the slot without disturbing the residue layer on either side of the row.

The timing of crimping matters, and it's specific. Crimp too early, before the cover crop has reached the flowering stage, and the plants may regrow from the base rather than dying cleanly. The vascular tissue of a young plant is resilient enough to recover from a single crimping pass. Crimp at or just after the flowering stage, when the plant has committed its energy to seed production and the stem tissue has begun to lignify, and the plant dies reliably without regrowth. For cereal rye, the most commonly used cover crop for crimping in temperate regions, the correct window is when the anthers are fully emerged and the pollen is visible. For legumes like hairy vetch, crimp when 50% of the plants are in full flower. These aren't suggestions. They're the biological thresholds that determine whether your crimping creates a dead mat or a temporarily flattened crop that bounces back and competes with your cash crop.

The roller-crimper itself doesn't need to be expensive or complex. A water-filled drum roller with a welded crimping pattern can be fabricated by any competent welder for a fraction of the cost of a commercial unit. The key specifications are a minimum weight of 500 kilograms per meter of roller width when filled, and a crimping pattern with chevrons or straight bars spaced at 15 to 20 centimeters apart. Lighter rollers don't create enough pressure to reliably sever the vascular tissue. Wider spacing between crimping bars leaves sections of stem intact that can recover. These two specifications, weight and spacing, determine whether your crimper works or merely flattens.

For stewards who aren't yet ready to invest in a roller-crimper, a high-clearance contact herbicide application using a product with no soil residual activity can terminate the cover crop without tillage. This is not the preferred approach because it doesn't create the woven mat structure that a physical crimping produces, and it adds a chemical input that your SRS Biology score doesn't benefit from. But it's a significant improvement over tillage because it leaves the root system intact, preserves the fungal connections, and allows the residue to fall and decompose on the surface rather than being buried. If you're using this approach as a transitional step, use the lowest effective rate of a contact-only product, apply in the morning when the cover crop canopy is dry and the stomata are open, and follow it with a light pass of a rolling basket or packer to press the dying biomass flat against the soil surface. The goal is to get the residue horizontal and in contact with the soil as quickly as possible so it begins functioning as armor immediately.

The cash crop goes in through the mat. A no-till seeder with adequate down-pressure, typically 150 to 200 kilograms per row unit for heavy residue conditions, cuts through the crimped mat cleanly and places the seed at the target depth without disturbing the residue on either side. The seed germinates in a microclimate that's cooler, moister, and biologically richer than anything a bare, tilled seedbed can offer. The mat suppresses weed germination by blocking the light that weed seeds need to trigger germination. The decomposing residue feeds the fungal network that immediately begins colonizing the new crop's roots. The soil temperature is stable. The moisture is retained. The armor is doing every one of its three jobs simultaneously, from the moment the cash crop seed touches the soil.

This is what it means to turn a standing crop into a biological bandage. You're not just terminating a cover crop. You're building the next layer of your Soil Resonance Score while simultaneously planting your cash crop. The same operation that used to require a tillage pass, a herbicide application, and a separate planting run now happens in a single pass that improves your SRS rather than degrading it. That's not just a technique. That's the Fibonacci logic applied to field operations: one action that produces compounding benefits across multiple SRS horizons at the same time.

Your Field Protocol: Putting Armor to Work

Everything in this chapter points toward a single, non-negotiable commitment: the soil stays covered. The three functions of armor, thermal regulation, kinetic dissipation, and the fungal bridge, all depend on continuity. Partial cover gives you partial protection. Seasonal cover gives you seasonal benefits. But 365-day cover, maintained through the deliberate management of both Green Skin and Brown Shield in sequence, is what produces the compounding SRS improvement that moves you toward the Inversion Point.

Here are the specific actions to take now. Each one has a clear, measurable outcome tied directly to your Armor score and the three functions it serves.

The first action is to assess your current armor coverage today. Walk every field on your property and, in each one, crouch down at soil level and look across the surface. Count how many times in a 30-meter transect you can see a patch of bare soil larger than your hand. Record the number. Do this in five locations per field. Average the counts. A field with zero bare patches in your transects has an Armor score approaching 8 to 10. A field with five or more bare patches per transect has an Armor score of 3 or below. This is your baseline. Write it down with today's date next to your SRS Architecture score from Chapter 1. These two numbers together tell you the state of your cathedral's foundation and skin.

The second action is to measure your current residue depth. In the same fields, push a ruler straight down into the surface residue in ten random locations per field and record the compressed depth at each point. Calculate the average. If your average compressed residue depth is below five centimeters, your dew-point gradient is compromised, your kinetic dissipation is insufficient, and your fungal bridge carbon supply is below the threshold needed to maintain network density through the cold months. The target is a minimum of five centimeters of compressed residue across 100% of your field surface at all times. If you're below this threshold, identify which fields are most critical based on their Architecture score. Fields with the highest Architecture scores have the most to lose from inadequate armor. Prioritize those fields first.

The third action is to design your cover crop termination approach for the next planting window. If you have a roller-crimper or access to one, plan your crimping timing around the species-specific flowering thresholds described above. Mark those dates on your calendar now, before the cover crop is in the ground, so you're not making the decision under planting pressure. If you don't have a roller-crimper, source or fabricate one before the next season. The specifications are in this chapter. The cost of fabrication is recoverable in a single season of improved infiltration and reduced irrigation. If fabrication isn't possible this season, plan your contact herbicide termination using the morning application protocol above, and commit to pressing the residue flat with a packer within 48 hours of application.

The fourth action is to close your largest armor gap. Every farm has one: a period in the calendar year when the soil is consistently bare or under-covered. For most farms in temperate regions, this gap occurs in the six to ten weeks between summer cash crop harvest and autumn cover crop establishment. Identify your specific gap window. Calculate how many days of bare soil you're currently accepting in that window. Then design a bridging strategy: either a faster-establishing cover crop species that can be seeded immediately after harvest, such as buckwheat in warm conditions or a fast-establishing cereal in cool conditions, or a residue management approach that leaves enough standing stubble height and spread biomass to maintain five centimeters of compressed cover through the gap period. The target is to reduce your largest bare-soil window to under 14 days. That's the biological threshold below which your fungal network can survive the interruption without significant thread die-off.

The fifth action is the aesthetic one, and don't underestimate it. Identify one person in your farming network, a neighbor, a family member, an agronomist, whose opinion about the appearance of your fields influences your management decisions. Have a direct conversation with them about what a covered, "messy" field is actually doing for your SRS. Show them the data from your Slake Test. Show them the temperature gradient readings from Chapter 3. Show them the fungal thread counts from Chapter 4. The aesthetic shift from "clean rows" to "covered land" is one of the hardest changes in regenerative practice because it runs against decades of cultural conditioning. But it's also one of the most consequential. A steward who has fully made that aesthetic shift manages their land differently in a hundred small decisions every season. They leave residue instead of baling it. They let the cover crop flower before crimping instead of terminating early. They see the brown mat and feel satisfaction instead of embarrassment. That shift in how you see your land changes what your land can become.

The Fibonacci Pulse can't beat through naked soil. The Silent Architecture you've been rebuilding since Chapter 1, the fungal network you've been protecting since Chapter 4, the Circadian Inhale you've been timing since Chapter 3, all of it depends on the skin being intact. Armor is the second weight in your SRS for exactly this reason. It's the protector of everything that came before it in the sequence. Without it, the cathedral stands open to the weather. With it, the cathedral breathes, the pulse compounds, and the system starts doing work you used to have to pay for.

The soil doesn't need a clean surface. It needs a covered one. The moment you can look at a thick mat of brown residue and a tangle of living green cover and see it for what it is, a self-regulating thermal system, a passive irrigation mechanism, a fungal food source, and a kinetic shock absorber all operating simultaneously, you've made the shift from mechanic to alchemist. You're no longer managing parts. You're conducting the whole system. And the whole system, when its skin is intact, compounds in ways that no individual intervention ever could.

Chapter 6: Diversity as Resonance: The Symphony of the Field

The Ancestor Speaks:

"My precious child, come, sit beside me by the hearth. The fire dances, and its flickering light casts long shadows that stretch and shrink, never quite the same. It is like the song of our people – always changing, always rich. Can you hear the many voices rising and falling, a chorus of laughter, a melody of sorrow, the steady rhythm of our beating hearts? This is life, in its fullness.

"There was a time, not so long ago in the long memory of the Earth, when the new ways came. They saw a field, and in their minds, they heard only one note. A single, loud, unwavering sound. 'This,' they declared, 'is order. This is efficiency.' They planted their fields with only one kind of plant, one shade of green, one voice. They called it 'progress.' Can you feel the emptiness of that single, droning note, a lonely sound that echoes into a vast, silent space? It promised strength but delivered only fragility. It was the ***monoculture of the mind****, spreading its flat, barren silence across the land. It was the* ***Linear Void*** *simplifying the sacred.*

"They did not understand. A single note, held long and loud, does not make music. It calls to those who would feast upon its isolation. The pest, the blight, they are not enemies, my child.

*They are but sharp **tuning forks**, tapping the hull of a ship that is off-key. They heard the lonely, stressed signal of that single note, and they came, not for malice, but for a meal promised by a cry for help. For a field broadcasting on only one frequency is an open invitation, a vulnerability no living system can long sustain.*

*"But we knew. We understood that the Earth sings a **Symphony**. Not a single note, but a rich, complex chord, many voices intertwined, each playing its part. Look around you, even now, at the wild edge of the woods. See the many different leaves, the tall grass, the low-lying herbs, each growing side by side. Each of these plants, my child, speaks its own language through its roots. It sends out its own unique message, a sweet invitation to a different kind of helper, a different kind of **Great Weaver** thread, a different kind of protector.*

*"It is like our village. We have the storyteller, the weaver, the hunter, the healer, the gatherer. Each has their own gift, their own song. And when all these voices sing together, they create a harmony that is strong, resilient, and beautiful. The land is no different. When many kinds of plants grow together, each playing its distinct role – one reaching deep, one fixing the breath of the sky, one building rich compost, one offering sweet nectar to the sky dwellers – they create a **biological chord** so rich, so full of overlapping signals, that no single threat can easily find its target. The sheer complexity, the **Resonance**, makes the enemy confused, makes the sickness lose its way. It is the wisdom of the **Rule of Five**, written not by human hand, but by the Earth's own ancient intelligence.*

*"This is not about adding more; it is about restoring the richness. It is about understanding that true strength comes from relationship, from the intricate dance of many different lives interwoven. It makes the **Fibonacci Pulse** beat not just faster, but deeper, richer, more vibrant.*

*"Now, close your eyes. Can you hear it? The faint, beautiful **Symphony of the Field**? Feel the comfort, the strength, the sheer joy of that diverse chorus. The **Seventh-Generation** longs to hear this song, to live in its harmony. Will you be the one to tune the instruments, to invite all the voices back to the choir? Feel the joy of that abundance, that deep, resonant peace that comes from a world fully alive and in tune."*

Meditation: Many voices, one field

Diversity as Resonance

" The single crop, so thin, so frail,
A lone note ringing, where the predator is caught.
But a vibrant choir, roots entwined and deep,
Builds a strong fortress where no discord can creep.
Tune the whole field, till harmony is kept…"

Diversity as Resonance: The Symphony of the Field

A single note played on a piano isn't music. It's a sound. You can hold it down for as long as you like, and it will never become a song. Add a second note and you have an interval. Add a third and something changes. The three notes start to talk to each other. They create tension or resolution, brightness or shadow. They become a chord. And a chord is something a single note can never be: a system with more meaning than the sum of its parts.

Your field works the same way.

A monoculture is a single note. It's one species, one root architecture, one chemical signature pulsing through the soil. It's loud in one frequency and completely silent in every other. And that silence is the problem. Not because silence is bad in a philosophical sense, but because the biological world reads frequency the way a radio reads signal. A field broadcasting on one channel is easy to find, easy to exploit, and impossible to defend. Every pest, every pathogen, every stress event that's tuned to that frequency hits the entire system at once. There's nowhere to hide. There's no chord to get lost in.

This chapter is about building the chord. It's about understanding why diversity isn't a nice addition to a regenerative system. It's the signal architecture the whole system runs on. And it carries a Fibonacci weight of 3 in your Soil Resonance Score for a reason that goes deeper than species counts or cover crop aesthetics.

The Monoculture of the Mind

Before we can talk about what diversity does in the soil, we need to talk about what uniformity does to the mind. Because the monoculture didn't start in the field. It started in the way we were taught to see productivity.

The industrial model of agriculture is built on one idea: simplify to maximize. Remove the variables. Control the inputs. Standardize the output. If one thing grows well here, grow only that thing. If one chemical solves one problem, apply it everywhere that problem appears. The logic is clean. The math is addition. And for a short window of time, the yields agreed with the logic, which is why the model spread so quickly and so completely across so much of the world's farmland.

But the simplification created something that the yield numbers never showed. It created a frequency problem.

Every plant species exhales a specific chemical signature through its roots. We touched on this in Chapter 4 when we discussed how root exudates act as a dinner bell for specific microbial communities in the rhizosphere. Different plants produce different exudate profiles. Different profiles recruit different microbes. Different microbes perform different functions: nitrogen fixation, phosphorus solubilization, disease suppression, aggregate formation. A field with one plant species has one exudate profile. It's feeding one narrow slice of the microbial community and leaving the rest of the underground economy unemployed. The biological diversity of the soil collapses to match the biological diversity of the plants above it. The chord reduces to a single note. The underground city loses most of its population.

When that happens, the soil stops being a self-regulating system and starts being a substrate. It can no longer buffer against stress because the buffering capacity of a soil is a function of its biological diversity. A soil with 200 species of bacteria, 50 species of fungi, and a full complement of protozoa, nematodes, and arthropods can respond to almost any perturbation. When a pathogen arrives, there are already organisms present that prey on it or compete with it. When a nutrient becomes temporarily unavailable in one form, there are organisms that can convert it to another form. When the weather turns extreme, the diversity of metabolic strategies in the soil community means some fraction of it always survives and maintains function. This is biological resilience, and it's built entirely from diversity. A monoculture soil, dominated by the narrow microbial community that one plant species supports, has almost none of it.

The pest pressure that follows a monoculture isn't bad luck. It's physics. A stressed, biologically impoverished system broadcasting a single chemical frequency is sending out a distress signal that's audible to every organism that feeds on stressed plants. The aphid doesn't choose your field because it dislikes you. It finds your field because the signal is clear and unmasked. There's nothing else in the frequency space to create interference. The monoculture is a beacon.

Think about a composite scenario that captures this pattern well. Picture someone like Callum, a 39-year-old grain farmer managing 420 acres in a temperate region. He's been growing continuous winter wheat for eleven years on his best paddocks because the economics made sense and the machinery was already set up for it. In years one through four, the system worked reasonably well. By year seven, he was spending more on fungicides than he'd ever budgeted for. By year ten, his take-all root disease pressure had reached the point where some paddocks were barely covering their input costs. His agronomist recommended a break crop. Callum planted a single-species canola break and saw improvement for one season before the disease complex shifted to target the canola instead. He wasn't dealing with bad luck or bad genetics. He was dealing with a system that had been simplified to the point where it had no biological immune response left. The solution wasn't a different single note. It was the chord he'd never been taught to play.

The monoculture of the mind says: find the best single answer and apply it everywhere. The Fibonacci Pulse says: find the minimum number of relationships needed to create a self-regulating system, and build from there. That minimum number, as we'll see in the next section, is five.

There's one more dimension of the monoculture problem worth naming before we move on. When a field is biologically uniform, it's also temporally uniform. The same chemistry is being exhaled at the same time, in the same amounts, throughout the growing season. The soil biology has no variation to respond to. The mycorrhizal network, which as we saw in Chapter 4 is driven by the diversity and timing of root exudate signals, has nothing to navigate. It's not building a complex, multi-directional web. It's maintaining a simple, linear connection to one type of root. The network stays thin. The glomalin production stays low. The Architecture score stays suppressed. The Armor can be perfect and the Biology can be present, but if Diversity is at zero, the entire SRS formula is multiplying a non-zero number by a weight of 3 and getting almost nothing back from it. The chord needs at least three notes before it starts to carry the harmonic complexity that makes the system self-regulating. And three notes, in the Fibonacci sequence, is where the individual and the partner become a system.

The Rule of Five

In the Fibonacci sequence, 1 + 2 = 3. The individual note and its partner become a system. But 3 is still a simple chord. It has structure. It has function. It's better than one note. But it's fragile in the way that simple systems are fragile: one element fails and the whole chord loses its character. The sequence moves forward. 2 + 3 = 5. And five is where something qualitatively different happens.

Five is the number of the bloom, the shield, the pentagon, the starfish arm. As we noted in Chapter 5, nature consistently uses the geometry of five when it builds something designed to protect a living center. In your cover crop mix, five species is the minimum threshold for creating what we'll call a full biological chord: a system with enough functional diversity to be genuinely self-regulating rather than just marginally more complex than a monoculture.

The Rule of Five is specific. You need five species, but not just any five. You need five species that each occupy a distinct functional role in the system. A mix of five grass species is still a single-note chord played louder. The diversity has to be functional, meaning each species must contribute something the others don't. Here are the five roles, what they do for your SRS, and which species fill them most effectively.

The first role is the Nitrogen Fixer. This is your legume, the species that hosts rhizobial bacteria in root nodules and pulls atmospheric nitrogen directly into the soil food web. Without a nitrogen fixer in your mix, the system is dependent on either purchased nitrogen or the slow mineralization of existing organic matter. With one, you're running a free nitrogen cycle that directly supports your SRS Biology score. The best options for temperate systems are hairy vetch, crimson clover, and field peas. Hairy vetch is the most aggressive nitrogen fixer of the three and produces the highest biomass in cool conditions. Crimson clover establishes faster and works well in mixes where you need early ground cover. Field peas add diversity to the exudate profile and support a different community of rhizobial partners than the clovers. Use hairy vetch as your primary legume at 15 to 20% of your mix by seed rate, and include one of the others at 5 to 10% for functional redundancy. If one legume struggles in a particular season, the other maintains the nitrogen-fixing function.

The second role is the Deep Rooter. This is your biological drill, the species with a taproot architecture that penetrates compacted layers, creates biopores that persist long after the root itself has decomposed, and mines minerals from depths that fibrous-rooted species can't reach. The Deep Rooter directly supports your SRS Architecture score by creating vertical channels through any residual compaction. The best options are tillage radish, also called daikon radish, and chicory. Tillage radish grows a root that can reach 30 to 45 centimeters deep in a single season, then dies over winter and decomposes in place, leaving a clean channel lined with carbon-rich root tissue that earthworms and fungal threads immediately colonize. Chicory is slower but more persistent in mixed systems and produces a deeper, more durable biopore. Use tillage radish at 3 to 5% of your mix by seed rate. It's a small seed and a little goes a long way. Avoid exceeding

20% brassica content in any mix, as brassicas are poor mycorrhizal hosts and high concentrations can suppress the AMF network you've been building.

The third role is the Carbon Builder. This is your grass, the species with a dense fibrous root system that produces the highest volume of root biomass and above-ground residue. The Carbon Builder feeds the Brown Shield we covered in Chapter 5 and provides the high-carbon exudate profile that the Great Weaver thrives on. It directly supports your SRS Architecture score through glomalin production and your Armor score through residue volume. The best options are cereal rye, oats, and sorghum-sudan hybrid depending on your season. Cereal rye is the gold standard for cool-season mixes. It's cold-hardy, produces massive biomass, and has the deepest fibrous root system of any common cover crop. Oats establish faster and work better in short-window situations. Use cereal rye at 25 to 35% of your mix by seed rate in cool-season applications. It will provide the structural backbone of your cover crop mat when crimped.

The fourth role is the Pollinator Bridge. This is your flowering species, the one that opens the system upward to the above-ground food web by providing nectar and pollen for beneficial insects. The Pollinator Bridge is not just about bees. It's about the predatory and parasitic insects that are your primary biological pest defense. Parasitic wasps, hoverflies, lacewings, and ground beetles all depend on floral resources as adults, even when their larvae are predating pests. A field with no flowering species has no way to maintain populations of these beneficial insects between pest events. When the aphid arrives, there's nothing there to eat it. Buckwheat is the single most effective Pollinator Bridge species available for this purpose. It flowers within four to six weeks of seeding, produces nectar accessible to a wide range of insect mouthparts, and is also a strong mycorrhizal host, giving it double duty in your mix. Use buckwheat at 8 to 12% of your mix by seed rate. Phacelia is an excellent alternative in cool conditions where buckwheat establishes slowly.

The fifth role is the Mycorrhizal Host. This is the species specifically chosen for its ability to form dense, highly active AMF partnerships and produce the exudate profile that grows the most complex fungal network. Every species in your mix should ideally be a reasonable mycorrhizal host, but one species should be chosen specifically for this function. Sunflower is the strongest mycorrhizal host in the common cover crop palette. Its root system produces an exudate profile that recruits a wider diversity of AMF species than almost any other cover crop, and the fungal network it builds persists in the soil and benefits the following cash crop even after the sunflower itself has been terminated. Use sunflower at 3 to 5% of your mix by seed rate. It's a large seed and will dominate the stand if overseeded. Phacelia, as noted above, is also an excellent mycorrhizal host and can serve double duty as both Pollinator Bridge and Mycorrhizal Host in cool-season mixes where sunflower doesn't establish well.

A practical five-way mix for a temperate cool-season application might look like this: hairy vetch at 18% by seed rate, crimson clover at 8%, cereal rye at 30%, tillage radish at 4%, buckwheat at 10%, and sunflower at 4%, with the remaining 26% filled by oats for additional carbon and early establishment speed. That's technically seven species, but five distinct functional roles. The redundancy in the carbon-builder and nitrogen-fixer roles is intentional. It's biological insurance. If one species fails to establish due to weather or soil conditions, the function it was supposed to perform is still covered by its functional partner.

The SRS Diversity score for this mix would be assessed at the 8 to 9 range: more than five species, all five functional roles covered, both cool-season and warm-season representatives present if you include the sunflower, and a strong mycorrhizal host included. Multiply that score of 8 by the Fibonacci weight of 3 and you get 24 points added to your total SRS from the Diversity horizon alone. Compare that to the 3 points a monoculture cover crop scores on the same horizon, 1 multiplied by 3. The difference between a single-species cover crop and a well-designed five-way mix is 21 SRS points. That's not a marginal improvement. That's the difference between a system that's slowly building and one that's compounding.

There's one more dimension of the Rule of Five that goes beyond species selection: root architecture diversity. A full biological chord requires not just different species but different root depths operating simultaneously. Your cereal rye is working the top 15 centimeters. Your tillage radish is working the 15 to 45 centimeter zone. Your sunflower is reaching into the 30 to 60 centimeter zone.

Your legumes are nodulating in the top 20 centimeters but fixing nitrogen that diffuses through the profile. These different root depths are mining different mineral layers, exhaling different chemical signatures at different depths, and feeding different communities of microbes and fungi at different levels of the soil profile. The underground city has multiple floors, all occupied, all active, all contributing to the biological density that your SRS Biology score depends on. A single-species cover crop occupies one floor and leaves the rest of the building empty.

The Pest as a Tuning Fork

The aphid on your wheat isn't your enemy. The fungal lesion on your canola leaf isn't a failure. The caterpillar working through your brassica seedlings isn't a sign that your system is broken. All of them are diagnostic instruments. They're tuning forks tapping the hull of your biological ship and telling you exactly where the resonance is off.

This reframe is harder to accept than it sounds, because the emotional response to a pest event is immediate and visceral. You see damage. You feel loss. The instinct is to reach for something that will stop it fast. The Linear Void built an entire industry around that instinct, and it's an industry that profits from the instinct remaining unexamined. But every time you reach for a broad-spectrum pesticide in response to a pest signal, you're shooting the messenger and leaving the message unread. The pest returns next season, often in greater numbers, because the condition that attracted it hasn't changed. You've silenced the tuning fork without fixing the note it was pointing to.

Understanding pests as diagnostic tools requires understanding what they're actually responding to. Pest organisms, whether insects, pathogens, or nematodes, don't distribute themselves randomly across a landscape. They concentrate where the signal is strongest. And the signal they're reading is the chemical frequency of stressed, nutritionally imbalanced, or biologically impoverished plants.

A plant under nutritional stress produces a different exudate profile than a healthy plant. Its cell walls are thinner. Its sugar-to-nitrogen ratio is off. Its secondary metabolite production, the compounds that make it taste bad or toxic to insects, is reduced. From a pest organism's perspective, a nutritionally stressed plant is an easy meal and a reliable breeding site. The pest isn't causing the problem. It's finding the problem that already exists and exploiting it. The stressed plant is broadcasting a frequency that the pest is tuned to receive. The pest is the receiver. The stressed plant is the transmitter. Your job, when you see the pest, is to ask what's wrong with the transmitter, not how to destroy the receiver.

The specific diagnostic logic works like this. Aphids concentrate on plants that are high in soluble nitrogen and low in structural sugars. This happens when nitrogen is abundant in the soil water, typically from high soluble fertilizer applications, but the plant's carbon metabolism is disrupted, often from poor mycorrhizal connections or inadequate photosynthesis due to shading or compaction. When you see aphid pressure, you're seeing a signal that your SRS Biology score is low and your nitrogen chemistry is outpacing your carbon biology. The fix isn't an insecticide. It's reducing soluble nitrogen applications and increasing the mycorrhizal host density in your next cover crop mix to rebuild the carbon-mineral trade we covered in Chapter 4.

Fungal diseases like powdery mildew and rust concentrate on plants with poor air circulation and weakened immune responses. Plant immune responses are largely mediated by the same mycorrhizal partnerships that deliver minerals. A plant with a strong AMF connection produces more systemic resistance compounds than a plant operating in biological isolation. When you see fungal disease pressure, you're seeing a signal that your Architecture score is limiting air movement through the canopy and your Biology score is too low to support adequate plant immunity. The fix isn't a fungicide. It's improving your canopy diversity so that different plant heights and growth forms create natural air movement, and rebuilding the fungal network that supports plant immune function.

Root-feeding nematodes concentrate in soils with low predatory nematode populations, which in turn reflects low biological diversity in the soil food web. A soil with a full complement of biological diversity has a population of predatory nematodes, mites, and protozoa that keeps plant-parasitic nematode populations in check. When plant-parasitic nematodes become a problem, they're telling you that the predatory community has collapsed, which tells you that the food web diversity is insufficient, which tells you that your Diversity score is pulling your Biology score down. The fix is adding species to your cover crop mix that support a broader range of soil fauna, particularly brassicas in limited quantities for their glucosinolate chemistry, which suppresses nematode populations directly, and deep-rooted perennials that support the predatory mite communities that live in the deeper soil horizons.

The diversity-as-pest-defense mechanism works through what the source materials for this book call the Noise-Canceling effect. A diverse field is broadcasting on multiple frequencies simultaneously. The chemical signature of the marigold overlaps with the signature of the wheat. The volatile compounds from the buckwheat flowers interfere with the olfactory cues that aphids use to locate their host plants. The dense canopy of a mixed stand makes it physically harder for flying insects to navigate to a specific host. The predatory wasp that parasitizes the caterpillar is present because the buckwheat nectar fed it through the gap between pest events. The system is so full of biological signal that the specific distress frequency of any individual plant gets masked by the complexity of everything around it.

This is fundamentally different from pest control. Pest control is reactive. It responds to a pest event after the damage has begun. Noise-Canceling diversity is structural. It prevents the pest from finding the signal in the first place. A field with a high Diversity score doesn't have fewer pests because they've been killed. It has fewer pests because the conditions that attract them, the clear, unmasked distress signal of a stressed monoculture, have been replaced by a complex, multi-frequency biological chord that's too rich and varied to be exploited by any single organism.

The field edges deserve special attention here because they're the most powerful and most neglected diversity resource on most farms. The hedgerow, the pond edge, the forest fringe, the grassy waterway: these are the places where two ecological systems meet, and the biological diversity at those meeting points is consistently higher than in either system alone. In ecological terms, these meeting zones are called ecotones, and they function as the speakers of your farm, broadcasting the highest-frequency biological signal from the edges inward across the field. The parasitic wasps, the ground beetles, the hoverflies, and the spiders that are your most effective biological pest managers all live primarily in these edge habitats and move into the field center to hunt. A farm with rich, diverse field edges has a biological pest defense that operates across the entire field. A farm with bare, mowed, or chemically managed edges has eliminated the habitat that houses its own immune system.

Protect your field edges. Not just from machinery. From the temptation to tidy them. A hedgerow with multiple shrub species, native grasses, and flowering perennials is not an inefficiency. It's the highest-value biological infrastructure on your farm. It's housing the army that keeps your field in tune. Every meter of diverse edge habitat you protect is a meter of Noise-Canceling capacity extending into the field beside it. And every meter of edge you mow, spray, or remove is a meter of biological silence that the pest signal can fill uncontested.

The last diagnostic tool in this section is the weed. Weeds are the most misunderstood tuning fork in the entire system. A weed isn't a random invader. It's a pioneer species that's been recruited by the soil to fill a specific functional gap. The soil is always trying to move toward complexity. When a gap appears in the biological chord, either because a functional role isn't being filled by your managed species or because a particular soil condition is present that only certain plants can tolerate, the soil recruits a species from the seed bank that can fill that gap. The weed is the soil's own attempt at the Rule of Five.

Dock and plantain in compacted, wet areas are telling you your Architecture score is low and your drainage is compromised. Capeweed and flatweed in dry, bare patches are telling you your Armor score is failing and the surface biology is degraded. Thistles in compacted, low-diversity areas are telling you your Diversity score is insufficient and the soil is trying to create its own biological drilling with the thistle's deep taproot. Chickweed in high-nitrogen areas is telling you your nitrogen chemistry is outpacing your carbon biology. These aren't random. They're a map of your SRS weaknesses drawn by the soil itself, using the only language it has available: the plants it can grow from the seed bank without your help.

When you start reading weeds this way, the response to them changes completely. You stop trying to eliminate them and start trying to understand what they're replacing. If dock is telling you the soil is compacted and wet, the fix is improving your Architecture score through reduced tillage and adding a deep-rooted cover crop species to the mix. When the Architecture improves and the compaction resolves, the dock doesn't need to be there anymore. It ceases to be the necessary pioneer. It fades out of the stand as the managed species fill the functional gap it was occupying. You didn't kill it. You made it redundant by providing the diversity it was trying to create.

The Quick-Start Guide

Diversity carries a Fibonacci weight of 3 in your SRS because it's the Signal Multiplier of the entire system. It's the layer that converts the physical foundation of Architecture and the protective function of Armor into a living, self-regulating biological intelligence. Without it, you have a well-built, well-protected building with no occupants doing meaningful work inside it. With it, the building comes alive. The underground economy diversifies. The fungal network complexifies. The pest pressure drops. The soil reads its own deficiencies and recruits solutions. The system starts doing work you used to have to pay for.

The four actions below are sequenced deliberately. Each one builds on the previous and produces a measurable outcome you can track against your SRS.

The first action is to design your five-way cover crop mix for the next planting window. Write down one species for each of the five functional roles: Nitrogen Fixer, Deep Rooter, Carbon Builder, Pollinator Bridge, and Mycorrhizal Host. Use the species recommendations in Section 2 of this chapter. For each species, write down the seed rate as a percentage of your total mix by weight, making sure your total adds to 100%. Check that your brassica content doesn't exceed 20% of the total. Check that you have at least one strong mycorrhizal host, buckwheat, phacelia, or sunflower, included at a minimum of 8% by seed rate. Write this mix down as your baseline diversity protocol. Every cover crop you plant from this point forward starts from this five-role template. You can adjust species within roles based on season, availability, and cost, but the five functional roles must always be filled.

The second action is to walk your field edges this week and assess their biological value. Walk the full perimeter of your two most important fields. For each section of edge, record what's growing there: bare ground, mown grass, a single shrub species, or a diverse mix of grasses, shrubs, and forbs. Rate each section on a scale from 1 to 5 for diversity, where 1 is bare or monoculture and 5 is a multi-species mix with flowering plants present. Calculate the average edge diversity score for each field. Any section scoring 1 or 2 is a biological gap in your pest defense system. Identify one specific action for each low-scoring section: either stopping mowing and allowing natural succession, or actively seeding a mix of native grasses and flowering perennials suited to your region. The target is to have no edge section below a score of 3 within two growing seasons.

The third action is to read your current weed population as a diagnostic map. Walk your fields and note the three most dominant weed species in each area. For each species, look up its ecological indicator status, what soil condition it typically colonizes. A basic weed identification guide with ecological notes, or a quick search using the species name plus "soil indicator," will give you this information. Map the dominant weeds against your SRS scores for each area. Do the compaction-indicator weeds appear in the areas with your lowest Architecture scores? Do the high-nitrogen weeds appear in the areas where you've been applying the most soluble fertilizer? The correlation won't be perfect, but it will be clear enough to give you a spatial picture of where your SRS weaknesses are concentrated. Use this map to prioritize where your five-way cover crop mix gets planted first in the coming season.

The fourth action is to calculate the SRS impact of your current Diversity score and set a specific twelve-month target. Score your current Diversity horizon on a scale of 0 to 10 using the criteria from Chapter 2: number of species, functional role coverage, and presence of both cool-season and warm-season representatives. Multiply your current score by 3. Write that number down. Now score what your Diversity horizon would be after implementing the five-way mix and edge diversity actions above. Multiply that projected score by 3. The difference between those two numbers is the SRS improvement available to you from the Diversity horizon alone, without changing anything else in your management. For most practitioners currently running single-species or two-species cover crops, this difference is between 12 and 18 SRS points. That's a meaningful move toward the Inversion Point, and it comes entirely from adding biological complexity to a system that's already physically intact from the work you've done in the previous chapters.

The Fibonacci Pulse needs a chord to carry it. One note can't sustain a rhythm. Five notes, each playing a distinct role, each supporting the others, each filling a frequency that the others leave open, create something the soil can actually dance to. The Symphony of the Field isn't a metaphor. It's the measurable, scoreable, compounding reality of a system that's finally running on enough biological complexity to regulate itself. Your job is to provide the minimum number of instruments. The soil will conduct the rest.

Chapter 7: The Carbon Exhale: The Solar Pump

The Ancestor Speaks:

"My tender one, come, feel the warmth of my embrace. Let us stand now, not in the dim light of the evening, but in the vibrant heart of the day, under the glorious sun. Look upon this single, green leaf, catching the light like a prayer. Close your eyes, and feel the immense power pouring from the heavens, through the skin of your face, and into this fragile emerald hand.

"They taught us, those who forgot, that the plant was a mere vessel, taking from the soil until it yielded its fruit, then leaving it barren. They spoke of the earth as a vault, ever emptying, ever needing to be refilled by human hand. This was the myth of the taker, the lie of scarcity that fed the ***Linear Void****. It made us anxious, always chasing, always fearing depletion. Can you feel the weariness in that old story, the endless, grasping hunger?*

"But we knew otherwise. We saw the plant as a holy conduit, a ***Solar Pump****, drawing the fire of the distant sun, not just for itself, but for the dark, hungry world beneath its feet. This leaf, my child, is a generous heart, beating with the very breath of creation. It breathes in the sky's unseen essence and transforms it into liquid light, into sweet sugars. And then, with an act of profound, silent giving, it pumps that liquid light, that precious* ***Carbon Exhale****, directly into the soil, into the welcoming embrace of the* ***Great Weaver****.*

"Can you imagine it? Down through the roots, a river of golden energy, flowing to feed the hungry multitudes, the unseen architects, the tireless miners. It is the plant's ultimate investment, a sacred offering that nourishes the very foundations of the earth. The more it gives, the richer the dark world becomes. The more it gives, the more the **Great Weaver** *can unearth the hidden mineral treasures, returning them as gifts to the generous plant. This, my heart, is the unbroken circle, the silent promise, the very pulse of infinite abundance. This is how the* **Fibonacci Pulse** *sings its song of ever-increasing wealth, built on selfless giving, not on grasping.*

"But then, I remember the bitter taste of sorrow, of fear. When the fields lay bare, after the harvest. Brown. Silent. Exposed. They called it 'rest.' But the Earth did not rest. She starved. The Solar Pump, so vital, so generous, was switched off. The flow of liquid light stopped. The **Great Weaver** *recoiled, its threads hungry, its work stalled. The architecture, which took so long to build, began to crumble for lack of nourishment. Can you feel the chilling silence of that 'Brown Gap,' the deep ache in the soil when the generous flow ceases? It was like stopping the very heartbeat of the world.*

"Our ancestors understood that the land must always wear a living garment. Always a leaf, always a root, always the pump running, even through the cold seasons, even between the great harvests. For every moment the pump is running, every moment a leaf breathes, is a moment the earth grows richer, more resilient, more alive.

"Now, close your eyes once more. Feel the sun's warmth on your skin. Imagine it flowing through you, a current of pure energy. Then feel that current flowing from your own heart, down into the earth, a gift of pure love and connection. The **Seventh-Generation** *watches, not just for the grain in your bins, but for the richness in the soil. Did you keep the pump running? Did you honour the plant's great exhale? Feel the profound gratitude, the joyous connection that pulses through you when you become a partner in this unending cycle of generosity."*

Prelude: The long gift of green

The Carbon Exhale

" The green leaf breathes, a lung that drinks the sun,
Pumping sweet sugar till the soil's work is done.
Beware the brown gap, the fallow's hungry maw,
Where life's own engine stalls, defying nature's law.
Keep the green pulse beating, from first light to frost's last draw. "

The Carbon Exhale: The Solar Pump

Every living plant above your soil right now is doing something that no machine on earth can replicate. It's pulling energy from a star 150 million kilometres away, converting it into liquid sugar, and pumping a significant portion of that sugar directly underground as a gift to the biology living in your soil. It's doing this in real time, without invoices, without fuel, without any help from you. The sun is the source. The leaf is the collector. The root is the pump. And the soil is the bank that receives every deposit.

This is the **Carbon Exhale**. And it's the spark that ignites the entire engine.

Understanding this changes everything about how you see a living plant. It stops being a crop to be harvested and becomes a solar pump charging the long term capacity of your land. It stops being a product and becomes a mechanism. And once you see it that way, the most expensive mistake in regenerative farming—leaving fields bare and brown for months at a time—becomes impossible to justify.

The Myth of the Taker

The story we've been told about plants and soil is a story of extraction. The plant takes nitrogen. It takes phosphorus. It takes water. It takes until the harvest removes biomass and the bank account of the earth runs a little lower than it did before. This is the foundational assumption of the NPK model: soil is a vault, crops are withdrawals, and your job is to calculate the deficit and top it up before the next season begins.

That story is wrong. Not partially wrong. Structurally wrong.

Through photosynthesis, a plant does something extraordinary. It breathes in carbon dioxide, absorbs sunlight, and turns those inputs into glucose—high energy liquid carbon. Some of that carbon becomes stems, leaves, grain. But a healthy, unstressed plant does not keep all of it. In a functioning biological system, a plant releases a substantial portion of what it produces directly out of its roots as exudates—sugars, amino acids, enzymes, organic acids—a complex cocktail deliberately delivered into the rhizosphere, the living zone around the root.

That release is not a rounding error. It is the plant's primary investment strategy.

The plant isn't depleting soil. It's capitalising it.

Every day a living leaf intercepts sunlight, the root system below it runs a solar powered injection pump, pushing liquid wealth into the ground. The plant is the pump. The soil is the account. And the compounding is not magic—it is what happens when yesterday's investment becomes tomorrow's capacity.

Consider someone like **Elena**, a composite of the stewards who live in this tension. She's a mixed grazier, chemical free for years. Cover crops in winter. Careful rotation. Minimal disturbance on her core paddocks. She's doing the right things—and yet she still feels like she's supplementing a failing system instead of conducting a thriving one. Soil tests crawl upward by fractions. Input costs don't fall as much as they should. The system still feels tight.

The shift comes when she stops measuring what she adds to soil and starts measuring how long the land stays green.

She maps her **Days of Green** across the calendar year and realises her best paddocks have living cover for about 210 days out of 365. Her worst paddocks have fewer than 120. Over two years, the difference in organic matter trajectory between those paddocks is explained less by products and more by time. The pump ran for different lengths of the year. The account received deposits for different numbers of days. The math was always there. She just hadn't been reading it.

Seeing plants as solar pumps reframes your relationship with every green thing on your land. A weed pumping carbon while you decide what to plant next is not just a nuisance—it's a deposit. A cover crop standing longer than planned is not always a failure—it can be the pump running on borrowed time, building Biology and Architecture while you aren't looking.

The goal isn't to control every plant.

The goal is to keep the pump running as many days of the year as possible.

The Liquid Bridge

The physics behind this connects directly to everything you've built in the earlier chapters. The liquid carbon a plant exhales into the rhizosphere doesn't sit there. It calls in partners. It feeds microbes and fungi that begin doing what they evolved to do: mining mineral wealth locked in the soil's parent material. They dissolve phosphorus from rock faces. They release zinc from clay surfaces. They fix atmospheric nitrogen. They produce enzymes that unlock nutrients from organic matter.

In return for carbon, those organisms deliver a mineral package back to the root that no bag of fertiliser can replicate in either diversity or bioavailability.

This is the **Liquid Bridge**: carbon moving down, minerals moving up, fertility emerging as relationship.

And when the bridge is intact, it compounds.

Here is the compounding principle in plain terms:

What the plant built yesterday, added to what it builds today, becomes the capacity it has tomorrow.

More carbon feeds more biology. More biology delivers more minerals. More minerals grow a larger plant. A larger plant captures more sunlight. More sunlight produces more carbon. The loop feeds itself. Each cycle increases the size of the next.

The sequence is running. The pulse is beating. And the soil is getting richer with every turn.

Measuring the Pump: Brix

There is a simple way to assess the strength of your solar pump. It's called a **Brix reading**—a measure of sugar concentration in plant sap.

Brix is measured with a refractometer, a small handheld instrument that can be inexpensive and practical in the field. Place a few drops of sap on the prism, hold it to light, and read the scale. In many crops, a low reading is a signal of stress, limited mineral nutrition, or compromised biological partnership underground. Higher readings indicate stronger carbon metabolism, better mineral delivery, and a more active relationship between plant and soil.

High Brix plants are also, not coincidentally, more resistant to pest pressure. A plant with higher sugar concentration tends to build stronger tissue and more defensive compounds—the terpenes, phenols, and alkaloids that make plants harder to digest and less attractive to opportunistic pests. The same carbon metabolism that feeds soil also builds immunity in the plant above it.

Take Brix readings from five plants in your highest performing paddock and five in your lowest performing paddock. Record the averages. The difference between those numbers is a practical measure of the difference in pump output across your land—useful, repeatable, and visible.

The Liquid Carbon Pathway

The "dinner bell" analogy captures the basic mechanism of exudation, but the Liquid Carbon Pathway is more precise than that. Different plants release different exudate profiles, and those profiles recruit different microbial communities. A legume speaks one chemical language. A grass speaks another. The point isn't romance—it's function. Diversity above ground creates diversity below ground, and diversity below ground increases the range of services the soil can perform.

This is why Diversity matters in your scoring framework: a diverse plant community doesn't merely increase the amount of carbon flowing into soil—it increases the *types* of carbon, the kinds of partners recruited, and the functions performed.

A five species cover mix does not just work harder than a monoculture. It speaks in five languages at once. It recruits more specialised partners. It builds a system with more redundancy and more resilience.

The rhizosphere is the most biologically dense zone on your land. In healthy soil, microbial populations around roots can be orders of magnitude denser than the soil only centimetres away. That density is driven by carbon flow. The root is a sun underground. Everything orbits it.

Grazing: Keeping the Pump Running

One of the least discussed aspects of the Liquid Carbon Pathway is its relationship to grazing.

When an animal removes leaf material, the plant temporarily reduces exudate output—because it must rebuild its solar surface before it can invest heavily underground again. This is not failure. It's rhythm. The key variable is recovery time.

If grazing removes roughly half the leaf area and then the plant is given time to regrow, the pump returns to full output within days to a couple of weeks, depending on season and conditions. The soil community rebounds. The relationship continues.

But if grazing pressure is continuous—if the plant is repeatedly grazed before it recovers—the leaf never returns, exudation stays depressed, fungal partnerships weaken, and Biology scores fall. The pump runs at a fraction of capacity not because the sun stopped shining, but because the collector was kept too small.

Rotational grazing, with adequate recovery, is therefore not merely pasture management. It is pump management.

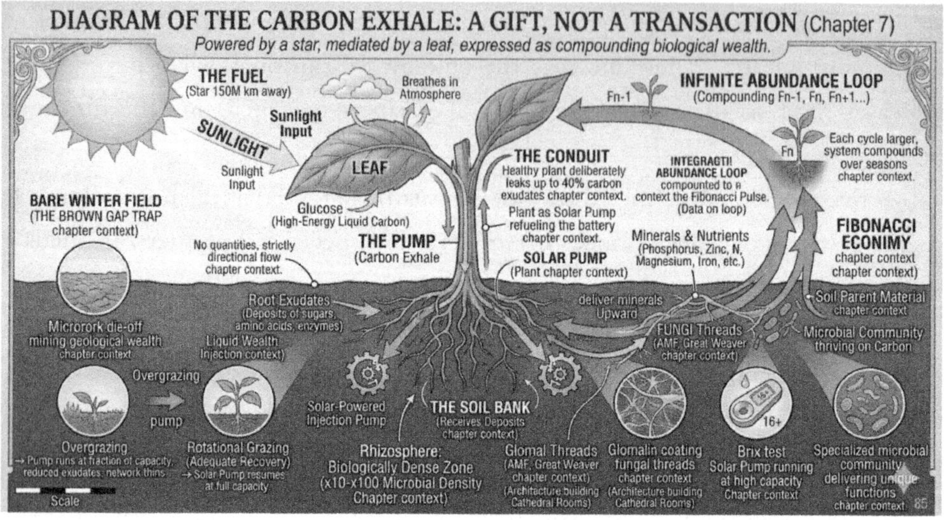

The Brown Gap Trap

There is a period in the year when many farms go quiet. The cash crop is harvested. The cover crop isn't established. The paddock turns brown and bare. The surface is exposed. This is treated as normal: the field at rest.

The soil doesn't rest during fallow.

It starves.

When the last living root is removed, the solar pump switches off. The carbon exhale stops. The rhizosphere community loses its food source. Mycorrhizal networks begin their countdown. Glomalin production slows. Aggregate formation stalls. Architecture begins to erode from the inside—not because the soil is lazy, but because the system was disconnected from its energy supply.

At the surface, bare ground faces full climatic stress. Heat denatures biological glues. Freeze thaw cycles fracture aggregates. Rain seals pores. The circadian moisture capture weakens without armor. Multiple failures occur at once because the fallow removes the one input the system cannot substitute: **a living leaf and root**.

The Brown Gap is expensive not only ecologically, but financially. After weeks of bare soil, fungal networks thin dramatically. When the next crop is planted, rebuilding those networks takes weeks. During that rebuilding window, plants operate in biological isolation, spending more energy on roots to compensate for what a network would have supplied.

The result appears later as a yield issue, a pest issue, a stress issue—when the real cause was a gap in continuity months earlier.

And the trap is self reinforcing weaker biology produces weaker plants; weaker plants invite pressure; pressure invites chemical intervention; intervention suppresses biology further; and the next Brown Gap hits an already compromised system even harder.

This is what it means to reset a compounding system.

Days of Green: the simplest lever you have

The solution is specific: maximise **Days of Green**.

Days of Green is simply the number of days in a calendar year that your land carries living leaf—capturing sunlight and pumping carbon below ground.

Every green day is a deposit. Every brown day is a pause—and, if prolonged, a withdrawal.

Calculate Days of Green for each paddock. Mark the days with active growth. Count them. Most conventional systems sit low. Regenerative systems rise as gaps close. Perennial systems approach continuous.

The difference between 150 Days of Green and 300 is not merely double. Because the system compounds, longer continuity produces a disproportionately larger return: denser biology, stronger networks, better structure, more resilience.

Eliminating the Brown Gap requires a practical operational commitment:

No paddock goes without living roots for more than fourteen days after harvest or grazing exit.

Within two weeks, establish something—anything—that keeps the pump running. A simple bridging species is infinitely better than bare ground. Complex mixes can come later. Continuity comes first.

Perennial species deserve special consideration because they remove the Brown Gap entirely where they are established. Their roots remain. Their exudate streams persist. Their fungal networks do not collapse each season. The system compounds for years without being reset.

Old pastures and undisturbed edges are rich not merely because they are old, but because the pump was never switched off.

Where you can establish perennials, do it. They pay dividends across multiple horizons simultaneously, year after year, with fewer interventions required.

Chapter recap and action steps

The Carbon Exhale is the spark that ignites the entire engine. Without it, structure loses its mason. Networks thin. The night breath has less to work with. Armor protects less.

Every chapter has been building toward this truth:

A living leaf above soil is the prerequisite for compounding.

The pump does not need your help to run. It needs your commitment to keep it running.

Over the next two weeks:

1. **Calculate Days of Green** for every paddock over the past 12 months. Identify the three lowest.
2. **Plan a cover-crop bridge** so those paddocks are replanted within fourteen days of harvest/exit.
3. **Take Brix readings** from high- and low-performing paddocks and record baselines.
4. **Establish one perennial strip** this season—small is fine. Mark it as a long-term reference area.
5. **Recalculate SRS** after closing your biggest gap and track whether Biology improves and Architecture holds steady across what used to be a bare period.

The soil doesn't need sunlight from you. It doesn't need carbon from you.

It needs a leaf.

Keep a leaf above it. Keep the pump running. And the system will do what living systems do when continuity is restored:

It will build tomorrow from what you allowed it to build today.

Chapter 8: The Horizon of the Void: Identifying the Machine

The Ancestor Speaks:

"My wise one, draw near. Place your hand upon your chest and feel the rhythm of your own heart. It is not always even, is it? Sometimes it quickens with joy, sometimes it slows with peace. It responds to the world around it. Now, close your eyes. Can you feel the subtle dis-ease, the faint discord that lingers even when you have tilled less, planted more, and guarded the earth with newfound love? That persistent whisper that says, 'Something is still wrong'?

"My mother, your great-great-grandmother, she felt it too. She saw the new rhythms brought by the iron beasts, by the clocks that commanded the day. She watched as the men, good men, earnest men, followed schedules written on paper, not in the wind or the soil. She called it the ***Mechanical Metronome****. It ticked with a relentless, unfeeling beat, ignoring the slow breath of the* ***Circadian Inhale****, deaf to the soft hum of the* ***Mycelial Internet****, blind to the gentle pulse of the* ***Carbon Exhale****. Can you feel the tension that arises when life's flowing river is forced into a rigid, straight canal?*

"They told us, 'This is progress. This is efficiency.' But she saw the cost. She saw the tractor's lights blazing in the deepest hours of night, when the earth should have been drawing her deepest breath, shattering the fragile channels of the ***Silent Architecture****. She saw the chemical mists falling when the* ***Great Weaver*** *was most active, poisoning the very threads of life. These acts were not born of malice, my child. They were born of blindness. Blindness to the true rhythms of the Earth. These were the* ***Hallucinations of the Void****: that we could conquer nature, substitute her wisdom with our own concoctions, and separate ourselves from the very ground that held us.*

"The elders warned of this. They said that when man ceases to listen to the Earth's heart, he imposes his own, colder beat. And the land, she resists. She grows weary. She loses her memory. The connections fray. The ***Fibonacci Pulse****, which longs to compound and expand, finds itself constantly reset, interrupted, forced back to a meager 'one,' again and again. Can you feel the deep fatigue of that constant struggle, the weariness of a system forever pushed against its own nature?*

"It is like trying to force a wild bird to sing a tune it does not know. Its song will be broken, its spirit dimmed. But the bird itself is not broken. The land is not broken. It is the metronome that is out of tune with the song.

"Now, open your eyes. Look at your fields, your land. You have learned to see the architecture, the armour, the exhale. Now, I ask you to look deeper. Can you see the invisible strings, the silent commands, the ingrained habits that are not of the Earth, but of the Machine? It does not shout; it whispers from old calendars, from inherited notions of 'clean' and 'efficient.' It is subtle, pervasive.

"Feel the powerful truth: once you see these invisible chains, they begin to loosen. Once you name the metronome, you can choose to dance to a different rhythm. The ***Seventh Generation*** *will thank you, not just for what you build, but for what you dared to dismantle. For the quiet courage to stop beating the drum of the Void, and to instead find the true, resonant pulse of the living Earth."*

Reflection: When systems forget

The Horizon of the Void

" The metronome's harsh tick, a hollow, frantic sound,
Born of three blind delusions, on shattered ground.
Reclaim your seeing from the steel, the gnawing debt,
And find the ancient math no clock could ever set.
Step from the failing machine; embrace the spiral yet. "

The Horizon of the Void: Identifying the Machine

There's a moment most regenerative practitioners know well. You're standing at the edge of a field you've been working hard to heal. The cover crops are in. The tillage has slowed. You've read the books, done the tests, watched the earthworm counts climb. And yet something still feels off. The system still requires more from you than it gives back. The inputs are lower than they used to be, but they're not gone. The yields are steadier, but they're not compounding. You're doing the right things, but you can't shake the feeling that something invisible is still pulling against you.

That something has a name. It's the Machine.

Not a tractor. Not a chemical company. Not any single piece of equipment or any single bad decision. The Machine is the entire operating system of the Linear Void, the invisible set of assumptions, schedules, habits, and hallucinations that were built into modern agriculture long before you arrived on the scene. It runs in the background of almost every farm decision, every field calendar, every cost-benefit calculation. And until you can see it clearly, you can't fully step out of it, no matter how many cover crop species you're planting or how carefully you're timing your grazing rotations.

This chapter is about seeing it clearly.

It's the most uncomfortable chapter in this book, and that's intentional. You can't leave a system you haven't clearly identified. The previous seven chapters have been building the architecture of the Fibonacci Pulse from the ground up. The Silent Architecture. The Math of Grace. The Circadian Inhale. The Mycelial Internet. The Living Armor. Diversity as Resonance. The Carbon Exhale. Each of those chapters gave you something to build. This chapter asks you to look at what's been tearing it down.

The Mechanical Metronome

A metronome doesn't listen. That's the whole point of it. It ticks at a fixed rate, indifferent to the music, indifferent to the room, indifferent to whether the instrument it's timing is a violin or a jackhammer. It imposes rhythm rather than finding it. And if you try to play a living piece of music to a metronome that's set to the wrong tempo, the result isn't music. It's friction. Every note lands slightly wrong. Every phrase fights the beat. The musician exhausts themselves trying to conform to a tempo that has nothing to do with the song.

This is industrial agriculture's relationship with the land.

The Mechanical Metronome is the factory schedule imposed on a biological system. It says the wheat goes in on the first of May because the planting calendar says so, not because the soil temperature has reached the biological threshold where the mycorrhizal network is active enough to immediately colonize the incoming root. It says the field gets sprayed on June the fifteenth because the spray contractor has a window, not because the crop's Brix reading indicates a biological vulnerability that actually warrants intervention. It says the harvest comes in September because the dryer is booked, not because the plant has finished its carbon exhale cycle and the soil is ready to receive the transition. The schedule is built around logistics, around equipment availability, around the requirements of the supply chain. The soil's biological clock doesn't appear anywhere on that schedule, because the Machine was designed before anyone knew the soil had one.

The Frequency Clash this creates is real and measurable. When the Circadian Inhale is at its peak draw between 10 PM and 4 AM, as we covered in Chapter 3, and a tractor is running a nighttime tillage pass through the field, the capillary channels that were actively drawing atmospheric nitrogen and condensing moisture downward are being physically shattered at the moment of their greatest productivity. The soil was trying to breathe. The Machine scheduled its most disruptive operation during the deepest breath. The result isn't just a missed opportunity. It's an active biological setback that shows up in your SRS Architecture score weeks later, when the aggregate structure that the disrupted inhale was supposed to be building simply isn't there.

The same clash happens in every season and at every scale. The Mycelial Internet, which we mapped in Chapter 4, runs on a four-hundred-million-year-old biological clock. It builds its most complex connections during the slow, undisturbed periods between major weather events. The Machine's spring tillage pass arrives like clockwork, every year, precisely when the fungal network is beginning its most active growth phase following winter dormancy. Six to eight weeks of network rebuilding, wiped out in a single afternoon. The timing isn't accidental. It's structural. The Machine was designed to operate on a calendar, and the calendar was designed around human logistics, not biological rhythms.

The bare field is the Mechanical Metronome's signature. A field stripped clean after harvest, smooth and dark and orderly, is the visual signature of a system that has fully substituted the factory schedule for the biological one. To the Machine, that bare field represents efficiency. The previous crop is gone. The next input cycle hasn't started yet. The field is ready to receive whatever the calendar says comes next. To the soil biology, that same field is a catastrophe unfolding in slow motion. The Solar Pump described in Chapter 7 has been switched off. The AMF threads are beginning their fourteen-day countdown to die-off. The glomalin on the aggregate surfaces is being degraded by the surface temperature climbing past 60 degrees Celsius with no armor to moderate it. The Circadian Inhale is weakening as the temperature gradient collapses. Every biological process that the previous seven chapters described is degrading simultaneously, and the Machine considers this normal because the Machine can't read any of those signals. It only reads the calendar.

To identify where the Mechanical Metronome is crushing the natural rhythm of your land, you need to do something the Machine never does: compare your operational schedule against the soil's biological clock. Pull out your farm calendar from the last twelve months. Write down every significant field operation: every tillage pass, every chemical application, every harvest, every planting event. Next to each operation, write the time of day it occurred and the approximate soil temperature at the time. Now ask three questions about each entry. Was this operation performed during the peak of the Circadian Inhale, between 9 PM and 5 AM? Was it performed on soil that was actively supporting a living plant community, meaning the Solar Pump was running? And was it performed within six weeks of a previous tillage or major disturbance, meaning the Mycelial Internet was still in recovery from the last disruption?

Any operation that scores "yes" on two or three of those questions is a high-friction point. It's a place where the Mechanical Metronome is directly clashing with the biological clock. It's a point where your SRS is being actively suppressed by a scheduling decision rather than a stewardship failure. These high-friction points are your first targets for change, and they're often the easiest wins available, because fixing them doesn't require a new input or a new technique. It just requires moving the same operation to a different time of day or a different point in the biological calendar.

The most common high-friction operations, based on the patterns described throughout this book's source material, are nighttime tillage runs scheduled to beat the heat of summer days, chemical applications timed for evening calm when the Circadian Inhale is at peak draw, and post-harvest bare periods that extend beyond the fourteen-day AMF survival window. Each of these can be rescheduled without any change to the operation itself. The tillage run moves to morning. The chemical application moves to the mid-morning window between 7 AM and 11 AM when the soil is transitioning from inhale to exhale and the capillary draw is reversing. The post-harvest cover crop establishment gets pre-planned so the seed goes in within seven days of harvest rather than six weeks later when the window of opportunity has already closed and the biological cost has already been paid.

The Mechanical Metronome isn't loud. That's what makes it so hard to identify. It doesn't announce itself. It just ticks, steadily, indifferently, setting the tempo for every decision on the farm. And every time the biological system tries to find its own rhythm, the metronome is already there, ticking a beat that has nothing to do with the song the soil is trying to play.

The Three Hallucinations

The Mechanical Metronome couldn't run without a belief system to justify it. Machines don't sustain themselves. They need operators who believe in what the machine is doing. And the Linear Void has produced three specific beliefs, three hallucinations, that make the Machine feel not just reasonable but necessary. These aren't fringe ideas. They're the foundational assumptions of industrial agriculture, and they're so deeply embedded in the culture of farming that most practitioners hold them without ever having consciously chosen to.

Seeing them clearly is one of the most disorienting experiences in the transition to resonant stewardship. Because once you see them, you can't unsee them. And once you can't unsee them, the Machine starts to look very different from the efficient, rational system it presents itself as.

The first hallucination is the Hallucination of Independence. This is the belief that a plant can be healthy without the soil. Not just without healthy soil, but without the biological relationships that make soil a living system in the first place. The hydroponic greenhouse is the purest expression of this hallucination: a plant suspended in a chemical solution, producing yield without any contact with the biological architecture we've been building throughout this book. The yield numbers are real. The efficiency metrics are impressive. But the plant is on life support. It has no mycorrhizal partners mining minerals from parent material. It has no rhizosphere community performing the biological functions that produce secondary metabolites, disease resistance, and mineral density. It's producing biomass, but it's not running the Fibonacci Pulse. It's a single note played at high volume with no chord underneath it.

In field agriculture, the Hallucination of Independence expresses itself more subtly. It shows up every time a farmer looks at a yellowing crop and reaches for a foliar spray rather than asking what biological relationship has broken down underground. It shows up in the assumption that a soil test result is a shopping list rather than a diagnostic of a failing delivery system. It shows up in the language of "inputs," the word itself carrying the assumption that fertility is something you bring to the soil from outside rather than something the soil generates from within when its biological relationships are intact. The Hallucination of Independence is the belief that the plant and the soil are separate systems that can be managed independently. The entire Fibonacci Pulse is a refutation of that belief. The plant and the soil are one system. The Carbon Exhale is the proof. The Liquid Bridge is the mechanism. The Mycelial Internet is the infrastructure. Independence is the hallucination. Entanglement is the reality.

The second hallucination is the Hallucination of Substitution. This is the belief that a bag of NPK can replace the Carbon Exhale. That a purchased biological inoculant can substitute for the four-hundred-million-year-old fungal network that tillage destroyed. That a synthetic hormone can replace the complex signaling chemistry of a living rhizosphere community. The Hallucination of Substitution is the belief that any biological process can be adequately replicated by a purchased product, if you just find the right product and apply it correctly.

The math of substitution is the math of entropy. In a straight-line system, energy is lost at every transfer. You put in 100 calories of fossil fuel to get 10 calories of food. The ratio gets worse, not better, with every passing season, because the biological system that should be generating fertility for free has been replaced by a purchased system that degrades the conditions needed for the biological system to recover. The fertilizer suppresses the mycorrhizal trade. The suppressed trade weakens the plant. The weakened plant produces less Carbon Exhale. Less Carbon Exhale means less glomalin production. Less glomalin means lower Architecture. Lower Architecture means less water retention. Less water retention means more irrigation. More irrigation means more salt accumulation.

More salt accumulation means more fertilizer needed to push nutrients past the salt barrier. The Leaking Bucket economy of substitution doesn't just fail to build the system. It actively creates the conditions that make the system harder to rebuild. Every "Quick Fix" is a subsidy paid to the Void to stay a little longer.

The specific financial signature of the Hallucination of Substitution is an input cost that never goes down. If your fertilizer, chemical, and irrigation expenditure per hectare has stayed flat or increased over the last five years despite your regenerative practices, the Hallucination of Substitution is still running in at least part of your system. A system that's genuinely building biological fertility should show a declining input cost over a three to five year horizon, not because you're cutting corners, but because the biological delivery systems are coming back online and doing work you used to have to pay for. Flat or rising input costs are the financial signature of a system that's substituting rather than building. The Hallucination is costing you money every season it runs.

The third hallucination is the most seductive of all. It's the Hallucination of Control. This is the belief that if you kill enough enemies, you will finally have peace. That if you spray enough herbicide, the weeds will eventually surrender. That if you apply enough fungicide, the disease pressure will eventually stop. That if you run enough tillage passes, the compaction will eventually be solved. The Hallucination of Control is the belief that the biological world is a collection of problems to be eliminated rather than a collection of feedback loops to be understood.

We saw this clearly in the pest diagnostic section of Chapter 6. The aphid isn't your enemy. It's a tuning fork. The dock weed isn't an invader. It's a pioneer species recruited by the soil to fill a functional gap your management created. The fungal disease isn't a random attack. It's a diagnostic signal pointing to low biological immunity in a plant operating without adequate mycorrhizal support. When you reach for a broad-spectrum control agent in response to any of these signals, you're not solving a problem. You're silencing a messenger and leaving the underlying condition intact. The message comes back louder next season, because the condition that generated it hasn't changed and now the biological community that was suppressing it has been further depleted by the control agent you used.

The Hallucination of Control is also the most physically expensive of the three. Broad-spectrum pesticides don't discriminate between the organisms they're targeting and the organisms that were suppressing those targets. A fungicide application doesn't just kill the pathogenic fungus on the leaf surface. It reduces the AMF population in the rhizosphere, the same population that was building your Architecture score and supporting your plant's immune function. An insecticide application doesn't just kill the aphid. It kills the parasitic wasp that was keeping the aphid population in check. The control agent solves the immediate visible problem while creating the biological conditions that guarantee a larger version of the same problem returns next season. This is not a failure of the product. It's the logical outcome of applying a linear solution to a recursive system. In the Spiral, there are no enemies. There are only feedback loops. And every feedback loop is telling you something specific about the state of your SRS.

Consider a composite scenario that captures all three hallucinations operating simultaneously. Picture someone like James, a 46-year-old mixed cropper who transitioned to no-till four years ago. He's eliminated most of his tillage. His cover crop mixes are diverse. But his input costs have barely moved, his pest pressure is still significant, and he's been adding a biological inoculant product to every planting run for three years without seeing the soil biology improvement he expected. James is running the Hallucination of Independence in his belief that the inoculant can substitute for the mycorrhizal network his previous tillage history destroyed. He's running the Hallucination of Substitution in his continued reliance on soluble phosphorus applications that suppress the very fungal partnerships the inoculant is trying to establish. And he's running the Hallucination of Control in his annual herbicide program, which is depleting the biological diversity in his soil food web faster than his cover crops can rebuild it. None of these three hallucinations is obvious to him, because each one looks like a rational, evidence-based response to a real problem. The inoculant is a real product with real science behind it. The phosphorus is genuinely deficient in his soil test. The weeds genuinely are competing with his cash crop. The hallucinations don't look like hallucinations from inside the Machine. They look like good management. That's what makes them so hard to leave.

The way out of each hallucination follows the same logic. For Independence, the exit is recognizing that the biological relationship is the product, not the input that substitutes for it. For Substitution, the exit is calculating the input cost trajectory over five years and asking whether the substitution is building toward biological independence or maintaining chemical dependency. For Control, the exit is reading the pest or weed signal diagnostically, asking what SRS horizon it's pointing to, and addressing that horizon rather than silencing the signal. None of these exits is comfortable. All of them require looking at the Machine clearly enough to see where you're still inside it.

Reclaiming the Horizon

Standing at the edge of the Void and refusing to jump is not a passive act. It requires a specific reorientation of what you're measuring, what you're optimizing for, and what you're willing to call success. The Linear Void offers a very seductive definition of efficiency: maximum output per unit of input, measured over the shortest possible time horizon. By that definition, a bare field is efficient. A monoculture is efficient. A chemical Quick Fix is efficient. The Machine scores well on every metric the Void considers worth measuring.

The Fibonacci Pulse offers a different metric entirely. Not efficiency, but resilience. Not maximum output today, but compounding output over time. Not the lowest input cost this season, but the declining input cost trajectory over five seasons as the biological system comes online and begins doing work you used to have to pay for. These two metrics produce completely different decisions in the field, and you can only optimize for one of them at a time. The moment you choose the Void's efficiency metric as your primary measure of success, you've stepped back inside the Machine, regardless of what cover crop species you're planting or how carefully you're managing your grazing rotation.

Reclaiming the horizon starts with a specific diagnostic practice: mapping your friction points. Friction, in the context of the Fibonacci Pulse, is any place on your land where the biological system is fighting the management system. It shows up as persistent pest pressure in a specific paddock that doesn't respond to the same biological management working everywhere else. It shows up as an area of chronic compaction that keeps returning despite reduced tillage. It shows up as a section of fence line where the soil is visibly darker and more aggregated than the managed land five meters away, telling you that the undisturbed biology just outside your management zone is doing something your management zone can't replicate. These friction points are the places where the Mechanical Metronome is loudest. They're the places where the Machine's schedule and the soil's biological clock are in the most direct conflict.

Walk your land specifically looking for friction this week. You're not looking for problems in the conventional sense. You're looking for places where you spend the most money, the most diesel, and the most management attention relative to the biological output you receive. High spending, high effort, low biological return is the signature of a friction point. It's the signature of a place where the Void is strongest and the Pulse is weakest. Mark those points on a sketch map of your property. They are your priority zones for the elimination of mechanical interventions, and they're also the places where your SRS Architecture score is most likely to be suppressed by something the Machine is still doing.

The Observer Effect, drawn from quantum physics, is relevant here in a way that goes beyond metaphor. In quantum mechanics, the act of observation changes the behavior of the observed system. An electron exists in a superposition of possible states until it's observed, at which point the wave of possibilities collapses into a single reality. Your land operates on an analogous principle. A paddock that you approach with the intent of the Fibonacci Pulse, looking for biological architecture, reading the weed population as a diagnostic map, measuring the Slake Test result as a structural assessment, is a paddock you're interacting with differently than a paddock you approach looking for problems to eliminate. The attention itself changes the quality of the decisions you make. The decisions change the trajectory of the SRS. The SRS trajectory changes the biological reality of the paddock. The Observer is part of the system being observed.

This isn't mysticism. It's the practical consequence of the fact that every field decision you make is filtered through the lens you're using to see the field. A steward using the Machine's lens sees bare soil and thinks "ready." A steward using the Pulse's lens sees bare soil and thinks "emergency." The same physical reality produces a completely different decision, and that decision compounds over seasons into a completely different landscape. The lens you choose is the most consequential management decision you make all year, and it costs nothing to change.

Reclaiming the horizon also means being specific about which mechanical interventions you're willing to eliminate, and in what order. The temptation is to try to change everything at once, to eliminate all tillage, all synthetic inputs, and all monocultures in a single season. This is the Hallucination of Control applied to the transition itself. The biological system can only rebuild as fast as its foundational layers allow. If you eliminate tillage before you have a cover crop establishment system that can manage weed pressure without it, you'll face a weed problem that drives you back to tillage within two seasons. If you eliminate all synthetic inputs before your biological delivery systems are functional enough to replace them, you'll face a yield drop that makes the transition economically unsustainable. The sequence matters as much as the direction.

The Fibonacci logic of the transition is this: address the highest-weighted SRS horizon first, and let each improvement create the conditions for the next one. Architecture before Armor. Armor before Diversity. Diversity before Biology. Biology before Chemistry. This is the sequence the SRS formula encodes, and it's the sequence that produces the most stable, compounding transition rather than the most dramatic short-term change. A transition that follows this sequence looks slow in the first two years and accelerates sharply in years three and four as the compounding kicks in. A transition that ignores the sequence and tries to change everything simultaneously often produces a chaotic middle period, two to three years of inconsistent results, that drives practitioners back to the Machine before the biological system has had time to reach the Inversion Point.

The concept of the Quantum Ancestor, which the next chapters will develop in full, is already present in the act of reclaiming the horizon. Every intervention you choose to eliminate this season is a decision that compounds forward through time. The mechanical compaction you avoid this autumn is a pore space that the

Mycelial Internet will colonize this winter, producing glomalin that will still be in your soil architecture seven years from now. The bare ground you cover with a fast-establishing buckwheat crop this week is a Solar Pump that will run for sixty days, producing carbon deposits that will be expressed as aggregate stability and fungal network density long after the buckwheat itself has been crimped and composted back into the armor layer. Small acts of refusal, the decision not to run the night tillage pass, the decision not to apply the insurance fungicide when the Brix reading doesn't warrant it, the decision to leave the hedgerow untouched rather than mowing it for tidiness, these compound in the same way the Fibonacci sequence compounds. The early numbers are small. The ratio is constant. The system builds on itself. And the land that receives the benefit of those compounding decisions will be measurably, demonstrably different from the land that didn't, not in ten years, but in three.

The Void is not your enemy. It's a system that was built by people trying to solve real problems under real constraints, using the best mathematics available to them at the time. The math of addition solved the immediate problem of feeding a growing population. It did so by borrowing against the biological capital of the soil, a debt that's now coming due across millions of hectares of degraded farmland worldwide. Recognizing the Void isn't about blame. It's about seeing the system clearly enough to step outside it deliberately, one high-friction intervention at a time, and let the Fibonacci Pulse fill the space the Machine leaves behind.

Your Audit: Four Actions Before the Next Season

The chapter you've just read is only useful if it changes something you do on the ground. The Machine is most powerful when it remains invisible. The moment you can see it operating in your specific context, on your specific land, in your specific seasonal calendar, it loses most of its hold. These four actions are designed to make the Machine visible and give you a concrete first step out of each of its primary mechanisms.

The first action is your diesel and chemical audit. Pull your input records for the last two full seasons. Create a simple table with three columns: the input or operation, the cost per hectare, and the SRS horizon it was intended to address. For chemical inputs, identify which SRS horizon each product was targeting: a fungicide targets Biology, a herbicide targets Diversity management, a soluble fertilizer targets Nutrients. For mechanical operations, identify which SRS horizon each pass was intended to affect. Now sort the table by cost per hectare, highest to lowest. Look at the top five expenditures. For each one, ask a single question: is this input building the SRS horizon it's targeting, or is it substituting for a biological process that a higher SRS score in a foundational horizon would make unnecessary? A soluble phosphorus application that's compensating for a suppressed mycorrhizal network is substituting for Architecture and Biology, not building Nutrients. A herbicide program that's compensating for a monoculture cover crop system is substituting for Diversity, not managing weeds. This audit will show you, in financial terms, exactly how much the three hallucinations are costing you per season. Write that total down. It's your financial argument for the transition, and it's one you can return to every time the Machine's efficiency metrics make the status quo look rational.

The second action is your high-friction mapping exercise. Take a blank sketch map of your property. Walk every paddock this week with the specific intent of identifying friction points, the places where you spend the most management energy for the least biological return. Mark each friction point with a brief note describing the symptom: chronic compaction, persistent pest pressure, poor germination, low Brix readings, Slake Test failures. For each marked point, identify which of the three hallucinations is most likely driving the friction. Compaction that returns after every tillage pass is the Hallucination of Control: you're trying to mechanically solve a problem that has a biological cause. Persistent pest pressure that doesn't respond to increasing chemical rates is the Hallucination of Substitution: you're trying to substitute a chemical solution for the ecological balance that a higher Diversity score would provide. Poor germination in areas with adequate moisture is likely the Hallucination of Independence: the biological community that supports seedling establishment has been depleted and the plant is operating without its biological partners. This map is your transition priority list.

The highest-friction areas are where the Machine is most active and where the Fibonacci Pulse has the most ground to recover.

The third action is to choose one mechanical intervention to eliminate this season. Not reduce. Eliminate. Completely remove one specific operation from your calendar for the coming season and replace it with a biological alternative. The most impactful single elimination for most practitioners is the post-harvest tillage pass that was previously used to manage cover crop residue or weed pressure. Replace it with a roller-crimper termination of the cover crop as described in Chapter 5, combined with a no-till cash crop establishment through the residue mat. This single change eliminates the Mycelial Internet disruption, maintains the Architecture score through the transition period, keeps the Brown Shield intact for the Circadian Inhale, and removes the primary high-friction operation from your seasonal calendar in a single decision. Calculate the direct cost saving of eliminating this pass: fuel, equipment wear, labor. Write it down next to the biological benefit: six to eight weeks of uninterrupted fungal network growth that would otherwise have been reset to zero. The financial case and the biological case point in the same direction. That alignment is what makes this the most accessible first elimination for most practitioners.

The fourth action is to set your SRS target for the horizon most suppressed by the Machine on your land. Return to your baseline SRS from Chapter 2. Identify the horizon with the largest gap between its current score and its theoretical maximum. For most farms still running significant mechanical intervention, this will be Architecture. For farms with adequate no-till but poor cover crop continuity, it will be Armor. For farms with good physical management but persistent monoculture cash crops, it will be Diversity. Take the Fibonacci weight for that horizon and multiply it by three. That number is the SRS improvement available to you from a three-point improvement in that single horizon. Write it down as your twelve-month target. Then identify the one specific action, from the chapters covering that horizon, that will most directly move that score. Not the most dramatic action. The most targeted one. The action that addresses the specific mechanism by which the Machine is suppressing that horizon on your land. This is your seasonal focus. Everything else stays stable. One horizon. One targeted action. Three points of improvement. Multiplied by the Fibonacci weight. Expressed as a measurable SRS gain by the same time next year.

The Machine runs on invisibility. It runs on the assumption that the way things have always been done is the way they must be done. The moment you can see the Mechanical Metronome ticking in your field calendar, the moment you can name the three hallucinations operating in your input decisions, the moment you can map the friction points where the Void is loudest and the Pulse is quietest, the Machine loses its power over your management decisions. It doesn't disappear. The cultural pressure to farm the way the neighbors farm, the economic pressure to use the inputs the supply chain makes most available, the aesthetic pressure to maintain the clean rows that generations of farming culture have celebrated as signs of competence, these pressures don't vanish when you can see the Machine clearly. But they stop being invisible. And invisible forces are the only ones you can't choose to resist.

The Fibonacci Pulse has been trying to beat through your land for as long as your land has been land. The Machine has been interrupting it, season by season, operation by operation, hallucination by hallucination. The horizon you're reclaiming isn't a new destination. It's the one the soil has been pointing toward all along. Your job is simply to stop standing in the way of it.

Chapter 9: The Quantum Horizon: The Observer's Power

The Ancestor Speaks:

"My soul-child, come to me now. It is the hour between night and dawn, when the veil between worlds is thinnest. We stand in the very heart of the oldest field, where our people have walked for longer than the rivers remember. The air is still, expectant. Close your eyes. Can you feel it? Not the cool earth, not the rising dew, but something deeper, a vast, humming silence that holds all possibilities within its breath?

"For generations, they taught us to look for what was broken. To see the cracked soil, the withered leaf, the hungry pest. They taught us to fix, to mend, to fight. It was the way of the Mechanical Metronome, always seeking the flaw, always trying to impose control. And what they sought, they found. Because what the eye seeks, the world provides. Can you feel the heavy weight of that endless searching for lack, that constant battle against what is seen as wrong?

"But we knew a deeper truth. We knew that the earth, like a dream not yet fully formed, exists in a thousand states at once. This field, even at its most broken, holds the memory of a lush forest, a thriving grassland. It is all here, my child, woven into the very fabric of its atoms, a

field of probability. And you, my beloved, you are not merely a farmer toiling in the sun. You are the Observer. Your gaze, your intent, your very presence, it is the hand that gently, profoundly, shapes that dream into waking reality.

"I remember my own elder, his hands gnarled like roots, holding a single, dried seed. He would gaze at it, not with sadness for what it was, but with fierce, quiet knowing for all it could be. He saw the mighty oak, the dancing field of grain, the rich, dark soil, all contained within that tiny husk. And by that pure, unwavering vision, he collapsed the wave of possibility into a single, fertile future. What he observed with such focused love, the earth hastened to manifest.

"For everything here, my child – the ancient rock, the smallest microbe, the soaring eagle, and your own beating heart – it is all entangled. What you do here, in this small patch of earth, ripples out, not just across the field, but through time itself. It is a biological entanglement, a promise reaching out from this moment to the far-off future. When you heal a single square of earth, when you pour your clear, loving intention into the soil, the signal travels. It whispers to every root, every fungal thread, every distant particle that has ever been connected to this place. It is a non-local healing, a quiet miracle witnessed by the unseen.

"They might call this 'spooky action at a distance.' We called it 'the truth of connection.' We knew that the resonance of your heart, the clarity of your vision, was as real a force upon the land as rain or sun. A confused mind brings chaos. A peaceful, purposeful heart brings harmony. The soil reads your frequency, my child. It responds to your coherent presence. It asks you: 'Which future do you truly observe for me?'

"Now, close your eyes. Feel the ground beneath you. Feel the vast, intricate web of life, reaching through the soil, through time, connecting you to every seed that has ever sprouted and every harvest that will ever be. Feel your own power, not to dominate, but to truly see, to truly intend. The Seventh Generation is already entangled with your choice, listening for the frequency of your heart. You are not just planting seeds. You are collapsing a future into being. Feel the profound, humbling, exhilarating power of that sacred responsibility, coursing through your very being."

Meditation: The observer's weight

The Quantum Horizon

" Your eye, the light that bids the wave descend,
The steward's deep intent, the field's true friend.
Entangle heart and mind with quantum grace,
Till root breaks unseen walls, and finds its rightful place.
The soil observes you. Meet its knowing gaze. "

The Quantum Horizon: The Observer's Power

You've spent eight chapters building something real. The cathedral has rooms. The mason is working. The fungal threads are weaving. The Solar Pump is running. The armor is holding the gradient. The chord is playing. You've moved from the Linear Void's flat, two-dimensional view of dirt into a three-dimensional, living, compounding system that runs on the same mathematical sequence as a galaxy arm and a sunflower head.

Now we go deeper.

Not deeper into the soil profile, though we'll go there too. Deeper into the physics of what's actually happening when a steward steps onto a field with intention and presence. Because there's a layer of this system that the previous chapters haven't fully named yet. It's the layer where biology meets quantum physics. Where the steward stops being a manager of parts and becomes something the science of the last century has only recently found language for: an Observer who collapses the wave.

This isn't a departure from the practical work of the Fibonacci Pulse. It's the completion of it. The soil isn't just a chemical system. It isn't even just a biological system. At the subatomic level, it's a field of probability. And the person standing on it, paying attention to it, holding a vision of what it can become, is part of that field. You always have been. You just haven't had the physics vocabulary to describe it.

Beyond the Visible Soil

Stand at the edge of your worst paddock. The one with the cracked surface and the thin, pale crop. The one that drinks the most inputs and gives the least back. Look at it through the lens you've been building throughout this book. You can see the collapsed architecture. You can read the weed population as a diagnostic map. You can feel the absence of the Circadian Inhale in the sealed, bare surface. You understand, now, what's missing and why.

But what you're seeing is still only the visible layer.

Beneath the chemistry, beneath the biology, beneath even the fungal threads and the glomalin bonds, the soil is made of atoms. And atoms are not the solid, billiard-ball objects that high school physics described. They're mostly empty space. They're clouds of probability, regions where a particle is more or less likely to be found, governed not by the laws of Newtonian mechanics but by the laws of quantum physics. At this level, the soil isn't a collection of things. It's a collection of fields. Overlapping, interacting, entangled fields of probability that extend beyond the physical boundaries of any single particle.

The Linear Void built its entire model on the Newtonian view. Separate particles. Separate causes. Separate effects. Add this mineral, get that result. Apply this chemical, produce that yield. The model is clean and logical and it works well enough at the scale of tractors and tonnes per hectare. But it misses the deeper operating system entirely. And when you're trying to understand why a soil managed with care and presence consistently outperforms a chemically identical soil managed with cold, mechanical distance, the Newtonian model has no answer. The quantum model does.

A soil that's being actively managed by a steward who understands its architecture, who walks it with attention, who reads its signals and responds to them, isn't just receiving different physical inputs than a neglected soil. It's in a different relationship with the person managing it. And at the quantum level, that relationship has physical consequences. The fields of probability that constitute the soil's subatomic structure are influenced by the fields of probability that constitute the steward's attention and intention. This isn't poetry. It's the logical extension of what quantum biology has been discovering since the 1990s: that living systems operate at the quantum level in ways that classical biology never accounted for, and that the boundary between the observer and the observed system is far less fixed than the Newtonian model assumed.

The soil you're standing on right now exists in a superposition of states. It's simultaneously the depleted, compacted, architecturally collapsed system the Machine made it, and the rich, porous, biologically complex cathedral it has the potential to become. Both states are real. Both are present in the quantum probability field of the soil. The question isn't which one is true. The question is which one gets collapsed into physical reality by the act of observation and intention. That's the work of this chapter. That's the role of the Quantum Observer.

The Observer Effect

In 1927, Werner Heisenberg described something that still makes physicists uncomfortable a century later. He showed that the act of measuring a quantum particle changes the particle's behavior. Before you observe an electron, it exists in a superposition, a simultaneous cloud of all its possible positions and states. The moment you observe it, the wave function collapses. The cloud of possibilities resolves into a single, specific reality. The observation didn't just record what was there. It determined what was there.

This is the Observer Effect. And it's not a quirk of laboratory equipment. It's a fundamental feature of how reality operates at the quantum level.

Now consider what this means for a steward standing in a field.

Your land exists in a state of superposition. It's simultaneously a dying dust-bowl and a lush, functioning ecosystem. Both states are encoded in its quantum probability field. The compacted, glomalin-depleted, biologically impoverished state is real. The porous, fungal-thread-rich, architecturally intact cathedral is equally real, as a potential, as a probability, as a wave that hasn't yet been collapsed into physical expression. Every time you walk onto your land with the intent of the Fibonacci Pulse, with the specific, informed attention of a steward who knows what the Silent Architecture looks like and holds the vision of it even in broken ground, you're doing something that's physically analogous to what the physicist does when they observe the electron. You're collapsing the wave. You're pulling one of those superimposed states toward physical reality through the act of focused, intentional observation.

This is why the shift from Mechanic to Alchemist, from Extractor to Steward, from Victim to Observer, described in the opening of this book, isn't just a philosophical reframing. It's a functional change in your role within the system you're managing. A mechanic fixes parts in isolation. A mechanic's attention is focused on what's broken, on what's missing, on what needs to be added or removed. That attention is a form of observation too. And it collapses the wave toward the broken state, because that's the state it's looking for and finding. An Observer who walks the same land with the vision of the Great Weaver intact beneath the surface, who sees the potential architecture in a cracked and compacted field, who holds the frequency of the Fibonacci Pulse even when the visible evidence is discouraging, is collapsing a different wave. They're pulling a different probability into physical expression.

Think about a hypothetical scenario that illustrates this. Picture someone like Miriam, a 44-year-old permaculture designer who took over a severely degraded 60-acre property three years ago. The previous owner had run continuous tillage for twenty years. The Slake Test failed immediately. The weed pressure was severe. The Brix readings were below 4 across the entire property. Every agronomist she consulted told her the same thing: the soil was too far gone for biological recovery to be economically viable within a reasonable timeframe. She should use a heavy chemical reset and start from scratch. Miriam didn't do that. Instead, she walked the property every morning for the first three months before doing anything else. Not to check for problems. Not to plan interventions. Just to observe. She looked for the places where the soil was trying to recover on its own. The corner where a volunteer clover had established. The fence line where the aggregate structure was slightly better than the surrounding paddock. The patch where, after rain, the water infiltrated instead of running off. She was mapping the probability field of her land. She was finding the places where the wave of recovery was already partially collapsed. Those became her starting points. Three years later, her Slake Test results are in the 6 to 7 range across 40 of the 60 acres. Her Brix readings average 11. Her input costs have dropped by 60%. The agronomists who told her it was impossible haven't asked how she did it. But the answer starts with the morning walks, before any seed was in the ground or any amendment was applied. She observed the potential into existence before she planted it.

The practical application of the Observer Effect isn't abstract. It's a specific daily practice. Walk your land not to audit its failures but to witness its potential. Look for the places where the Fibonacci Pulse is already beating, however faintly. The single earthworm in a paddock that otherwise shows none. The volunteer plant with a Brix reading two points higher than its neighbors. The patch of soil that holds together slightly longer in the Slake Test than the surrounding area. These aren't anomalies. They're the leading edge of the wave collapse. They're the probability field of your land expressing its potential in the most favorable micro-conditions available. Your job is to find them, observe them with full attention, and let that attention amplify the signal they're broadcasting into the rest of the system.

Your belief in the life of the soil provides the energetic frequency that allows that life to manifest. This isn't a motivational statement. It's the practical consequence of the Observer Effect applied to land stewardship. A steward who genuinely believes their soil is beyond recovery makes management decisions that confirm that belief. They don't do the Slake Test because they already know what it will show. They don't protect the Circadian Inhale because they don't believe the biology is there to benefit from it. They don't design a five-way cover crop mix because they're convinced the system is too degraded to respond. Their observation collapses the wave toward the depleted state, not through malice, but through the physics of focused attention landing on the worst-case scenario and treating it as the only scenario. The steward who holds the vision of the cathedral, even while standing in the ruin, makes different decisions. And those decisions, compounding over seasons, produce a different physical reality.

Biological Entanglement

In 1935, Albert Einstein, Boris Podolsky, and Nathan Rosen described a phenomenon in quantum mechanics that Einstein famously called "spooky action at a distance." Two particles that have interacted become entangled. What happens to one instantly affects the other, regardless of the distance between them. Measure the spin of one particle and the spin of its entangled partner resolves instantaneously, even if that partner is on the other side of the planet. No signal passes between them. No time elapses. The connection is non-local. It transcends the physical distance separating the two particles as if that distance doesn't exist.

Quantum entanglement isn't a theoretical curiosity anymore. It's been experimentally confirmed thousands of times. It's the basis of quantum computing and quantum cryptography. It's a real feature of how the physical world operates at the subatomic level.

Now consider what this means for a soil system where every component, the minerals, the fungal threads, the bacteria, the plant roots, and the steward standing above it, has been in continuous physical interaction for seasons or decades.

Every time a root tip contacts a mineral grain, an interaction occurs at the quantum level. Every time a fungal thread produces glomalin and coats a soil particle, an entanglement is formed between the biological system and the mineral substrate. Every time you press your hand into the soil, your body's own quantum fields interact with the fields of the soil beneath your palm. These interactions don't end when the physical contact ends. In a quantum system, interaction creates entanglement. And entanglement means that what happens to one part of the system is felt, instantly and non-locally, by every other part it has been entangled with.

This is why healing one square meter of soil sends a signal across the entire field. Not as a metaphor. As a physical consequence of the entangled field that a managed soil system represents. When you improve the Architecture score in one corner of a paddock, by protecting the pore space, building the glomalin, and establishing a living root, the quantum signal of that improvement propagates through the entangled field of the entire system. The microbes in the far corner of the paddock, connected through the Mycelial Internet and through the quantum entanglement of a shared biological history, respond to that signal. The mineral spin states in the bedrock, which have been in continuous interaction with the biological system above them for geological timescales, shift in response to the change in the biological field above them. The system is non-local. The healing is non-local. Your action in one place reaches everywhere the field extends.

The practical implication is specific and important. You don't need to fix everything simultaneously. The quantum field of your soil means that every improvement you make in any part of the system sends a signal to every other part. Start with your highest-SRS-potential area, the corner with the best existing architecture, the patch with the most fungal threads visible in the aggregate test, the section that already shows the strongest Brix readings. Pour your first season's effort into that area. Bring it to the highest SRS score you can achieve. Then watch what happens in the surrounding areas. The signal of that improvement propagates. The entangled field shifts. You'll see responses in areas you haven't physically touched yet, because the system is connected at a level that goes below the visible biology into the quantum substrate of the soil itself.

There's a second dimension of biological entanglement that the source materials for this book describe with precision: the entanglement between the mineral spin states in your soil's bedrock and the intent in your mind. The subatomic particles that make up the minerals in your soil have quantum spin states. These spin states are influenced by the electromagnetic fields of living biological systems, including the human nervous system. A steward who approaches their land with a coherent, focused electromagnetic field, the kind produced by a calm, attentive, intentional nervous system, is literally interacting with the quantum spin states of the minerals in the soil beneath their feet. The interaction is weak at the individual particle level. But across the trillions of mineral particles in a single hectare of soil, all of them entangled with the biological history of the land and with each other through that history, the cumulative effect of coherent human electromagnetic presence is not negligible.

This is why land managed with love and presence consistently produces more than land managed with cold, mechanical distance, even when the chemical inputs are identical. The Resonance is different. The steward is part of the quantum field of the system they're managing. Their internal state, their coherence or their anxiety, their vision of abundance or their expectation of failure, propagates through the entangled field of the soil as surely as the liquid carbon from a plant's root exhale propagates through the rhizosphere. The soil reads you. Not metaphorically. Physically. At the level where everything is fields and probabilities and entangled spin states rather than separate objects in empty space.

The practical discipline this calls for is specific. Before you walk onto your land, especially in the areas you're trying to heal, take a moment to shift your internal state deliberately. Not into false optimism. Into accurate observation of potential. There's a difference between pretending the compaction isn't there and holding the vision of the architecture that the compaction is temporarily suppressing. The first is delusion. The second is quantum stewardship. You're not ignoring the current state. You're choosing which probability you're going to amplify with your attention. You're deciding which wave you're going to collapse with your presence.

Quantum Tunneling of the Root

Hard-pan is the word that makes most regenerative practitioners go quiet. The dense, compacted layer that sits six to eighteen inches below the surface, impenetrable to water, impenetrable to roots, the layer that all the cover crops and all the reduced tillage and all the careful management seems unable to fully break through. You can improve everything above it and still watch your crops run out of moisture in a dry spell because the roots can't reach the water stored below the hard-pan barrier. It's the wall at the bottom of the cathedral that no biological mason seems able to dissolve.

At the Newtonian scale, hard-pan is a physical barrier. Particles packed so tightly together that water can't flow through them and roots can't penetrate them without applying more force than their growing tips can generate. The Linear Void's answer to hard-pan is mechanical: a deep ripper, a subsoiler, a chisel plow run deep enough to shatter the layer. It works, temporarily. The layer reforms within two to three seasons as the biological system that should be keeping it open remains too depleted to maintain the biopores the mechanical intervention created.

At the quantum scale, hard-pan looks different.

Quantum tunneling is the phenomenon where a particle passes through a barrier that classical physics says it can't cross. It doesn't go over the barrier or around it. It passes through it, by existing on the other side. This happens because at the quantum level, a particle's position isn't fixed. It's a probability distribution. There's always a non-zero probability that the particle will be found on the other side of any finite barrier. When the barrier is thin enough and the particle's energy is high enough, tunneling happens regularly. It's not a rare exception. It's a standard feature of quantum mechanics that underlies the functioning of your phone's transistors, the nuclear fusion in the sun, and the enzyme reactions in every living cell on earth.

In a soil system in a state of high resonance, where the Fibonacci Pulse is beating strongly through a rich biological community, the barriers between problem and solution become porous in a way that parallels quantum tunneling. Water molecules, moving through the probability field of a highly structured soil, find pathways through what appears at the macro scale to be an impenetrable clay layer. Root tips, guided by the chemical gradients produced by the mycorrhizal network, follow pathways through the hard-pan that were created by previous root decomposition, by earthworm channels, by the freeze-thaw cycling of water in biopores. These pathways are invisible at the surface. They look like a solid barrier from above. But at the scale where the biology operates, the barrier is porous with potential routes, and a root tip guided by a healthy, active fungal network finds those routes through a process that looks, at the macro scale, like penetration of an impenetrable barrier.

The key phrase is "in a state of high resonance." A root tip operating in biological isolation, in a soil where the Mycelial Internet has been severed and the rhizosphere community is thin, doesn't have the chemical guidance system that allows it to find the pathways through hard-pan. It hits the dense layer and stops, or turns sideways, or circulates in the upper profile until the plant's moisture and mineral supply runs out. But a root tip operating in a fully connected biological system, guided by the chemical signals of a dense and active mycorrhizal network, follows the quantum probability gradients of the soil's entangled field toward the pathways of least resistance. It doesn't force its way through the hard-pan. It finds the way through that the resonance of the system reveals.

This reframes the hard-pan problem completely. Hard-pan isn't primarily a mechanical problem requiring a mechanical solution. It's a resonance problem requiring a resonance solution. A soil with a high SRS Architecture score, a dense and active Mycelial Internet, and a continuous Carbon Exhale from living roots above it is a soil where the biological drilling described in Chapter 6, the tillage radish, the chicory, the deep-rooted perennials, operates at its maximum effectiveness. The deep rooters don't just push through the hard-pan mechanically. They follow the quantum probability pathways that a resonant soil reveals. They find the channels. They widen them. They line them with carbon-rich root tissue that earthworms colonize. They leave behind biopores that persist because the biological community maintaining them is dense enough to keep them open. The hard-pan doesn't break all at once. It becomes progressively more porous as the resonance of the system increases, exactly as the quantum tunneling probability increases as the energy of the particle increases relative to the barrier height.

The specific application for managing hard-pan without mechanical intervention follows directly from this understanding. The target isn't the hard-pan layer itself. The target is the SRS of the soil above it. Every point you add to your Architecture score in the upper profile increases the resonance of the system. Every meter of fungal thread you add to the network above the hard-pan increases the chemical guidance available to root tips approaching the barrier. Every species you add to your cover crop mix that includes a deep-rooted representative, tillage radish at 4% of your seed mix, chicory at 3%, alfalfa as a perennial component where the system allows, is adding a biological drilling force that operates through resonance rather than force. The combination of high-SRS soil above the hard-pan and diverse deep-rooted species pushing downward from within that high-SRS system is the most effective hard-pan management strategy available. It's slower than a subsoiler. But it's permanent. The biopores it creates stay open because the biological community maintaining them is continuously renewed by the living root system above them. The subsoiler's channels close within three seasons because the biological community that should keep them open was never rebuilt.

There's a specific diagnostic for assessing whether your hard-pan management strategy is working through resonance or still fighting friction. After your second season of deep-rooted cover crops in a no-till system, dig a profile pit in the area of your worst historical compaction. Go down to 45 centimeters. Look at the structure of the soil at the 20 to 35 centimeter depth, the typical hard-pan zone. If you see vertical channels lined with dark, carbon-rich material, the remnants of decomposed root tissue, and if those channels show white fungal threads running through them, your biological drilling is working. The tunneling is happening. The resonance is finding the pathways. If you see only horizontal layering with no vertical channels, your SRS in the upper profile is still too low to generate the resonance needed for the biological drilling to penetrate effectively. Go back to Architecture. Raise the Slake Test score. Increase the fungal thread density. Build the resonance first. The hard-pan will follow.

Your Practice: The Quantum Steward's Field Protocol

Everything in this chapter points toward a shift that's both internal and external. The external practices, the five-way cover crop mix, the armor management, the Circadian Inhale protection, the tillage reduction, these have been building through every previous chapter. They don't change. But this chapter adds a layer to every one of them: the quality of attention and intention you bring to each practice.

A steward who crimps a cover crop while mentally running through a list of problems is collapsing a different wave than a steward who crimps the same cover crop while holding the vision of the fungal network that will colonize the residue mat over the next sixty days. The physical action is identical. The quantum field interaction is not. The first steward is observing the problem. The second is observing the potential. The soil reads the difference.

Here are four specific practices. Each one is actionable, observable, and tied directly to the SRS framework you've been building throughout this book.

The first practice is the Intentional Walk. Once a week, walk one paddock on your property with a single, specific purpose: to find evidence of the Fibonacci Pulse beating. Not to audit problems. Not to plan interventions. To find the pulse. Look for a soil aggregate that holds together in your hand. A fungal thread visible when you break one open. A weed species that's less dominant than it was last season, telling you that the functional gap it was filling is being addressed by your managed diversity. A patch of soil that feels slightly different underfoot, slightly more yielding, slightly more alive than the surrounding area. Write down what you find. One observation per walk. Date it. This is your Quantum Observer's field journal. Over twelve months, it becomes a map of the probability field of your land, showing you where the wave of recovery is already collapsing into physical expression and where the potential is still waiting to be observed into reality.

The second practice is the Visualization of the Silent Architecture. When you stand in a broken area of your land, an area with low Slake Test scores, visible compaction, or poor cover, spend two minutes before any physical intervention doing this specific thing: visualize the cathedral. Not as wishful thinking. As an accurate description of what the quantum probability field of that soil already contains as potential. See the aggregates forming. See the glomalin coating the particles. See the fungal threads growing through the pore channels. See the water moving downward during the Circadian Inhale. You're not imagining something that isn't there. You're observing a state that's already present in the superposition of the soil's quantum field, and you're collapsing the wave toward it with the focused attention of a steward who knows exactly what healthy soil architecture looks like. Do this before every intervention in a broken area. The intervention that follows a clear visualization of the intended outcome is a different physical act than the same intervention performed without that focus. The quantum field of your intention is part of the system you're acting on.

The third practice is the Frequency of Gratitude. This one is the hardest to take seriously if you're coming from a purely mechanical view of land management, and it's the most important one in this chapter. Gratitude is not a sentiment. It's a coherent electromagnetic state produced by the human nervous system when it's in a condition of genuine appreciation. A coherent electromagnetic state produces a coherent quantum field interaction with the entangled system of the soil. Gratitude, expressed as a genuine acknowledgment of what the soil is already doing, the glomalin it's already producing, the carbon it's already anchoring, the mycorrhizal connections it's already maintaining despite everything the Machine has done to interrupt them, is the highest-frequency signal you can send into the entangled field of your land. It's not magic. It's physics. The soil has been trying to run the Fibonacci Pulse for as long as it's been soil. Acknowledging that effort, genuinely and specifically, is the act of an Observer who's paying enough attention to see what's actually happening at the quantum level. And that quality of attention changes the probability field of the system you're observing. Do this once a day. Specifically. Name one thing the soil is doing that you're grateful for. The Carbon Exhale from the cover crop you can see pushing through the crimped residue mat. The earthworm you found in a paddock where you've never found one before. The Slake Test clod that held for three minutes when last season it dissolved in thirty seconds. Specific, genuine acknowledgment. One thing. Every day. It costs nothing and it changes the frequency of your presence on the land in ways that compound through the entangled field of the entire system.

The fourth practice is the Non-Linear Mind. The Linear Void trained you to look for cause and effect. Apply this input, get that result. Run this operation, produce that outcome. The Fibonacci Pulse runs on resonance and reflection, not cause and effect. Resonance means the system responds to the frequency of what you bring to it, not just the physical inputs you apply. Reflection means the land is a mirror of your own internal state, your own biological coherence or incoherence, your own clarity or confusion about what you're trying to build. When a paddock isn't responding the way you expect, before you add another input or run another operation, ask a different question. Not "what is this soil lacking?" but "what frequency am I bringing to this paddock?" Not "what's wrong with this crop?" but "what does this crop's signal tell me about the resonance of the system it's growing in?" This shift in question type produces a different quality of observation. And a different quality of observation collapses a different wave. Start applying this question to your two lowest-SRS paddocks this week. Walk them with the question "what is the resonance here?" rather than "what is the problem here?" Write down what you notice when you look through that lens. The answers will surprise you, and they'll point toward interventions that the problem-focused lens would never have found.

The Quantum Horizon isn't a destination you reach after completing the physical work of the previous chapters. It's the layer of the Fibonacci Pulse that was always present beneath the physical practices, waiting for the steward to develop enough understanding of the system to recognize it. The soil has always been a quantum biological system. The Observer Effect has always been operating in your fields. The entanglement between your intention and your land's probability field has always been there. The quantum tunneling of roots through barriers that resonance reveals has always been the mechanism behind the biological drilling that works and the mechanical drilling that doesn't hold. None of this is new. The physics is a century old. The biology is four hundred million years old. What's new is the steward who can hold all of it simultaneously: the mud on their boots and the quantum field beneath their feet, the Slake Test result and the wave function it's measuring, the cover crop mix and the entangled biological intelligence it's feeding. That steward isn't just managing a farm. They're conducting the Source Code of the soil itself. And the Source Code, when it's finally free to run without interruption, compounds in ways that no linear model ever predicted and no mechanical metronome ever allowed.

The child who will stand on this land seven generations from now is already entangled with you. The carbon you anchor today, the glomalin you protect, the fungal network you refuse to sever, the wave of recovery you collapse into physical reality through the quality of your observation and intention, all of it propagates forward through the non-local field of biological entanglement that connects every steward to every future that their stewardship makes possible. You're not just a farmer. You're a Quantum Ancestor. And the frequency you bring to your land today is the signal that determines whether the world your descendants inherit is a desert or a cathedral.

Chapter 10: The Resonant Human: Becoming the Conductor

The Ancestor Speaks:

"My wise and tender one, we have walked a long path together. You have knelt in the soil, touched the hidden architecture, heard the night's breath, and seen the silver threads that bind all life. You have glimpsed the vast, generous exhale of the plant and dared to stare down the shadow of the Machine. Now, as the sun begins its slow descent, painting the sky with the hues of journeys completed and new horizons born, we come to the deepest truth of all.

"For too long, the 'old ways' of the **Linear Void** *taught you a small story about yourself. They said your life began and ended with your own breath, your own years. They called it the 'Me Horizon.' And from that cramped space, every seed you planted, every furrow you turned, every decision you made, was weighed against the short measure of your own days. 'Will I see the fruit?' 'Will I reap the reward?' This narrow vision, it made you weary, did it not? It clipped the wings of your generosity, muted the deepest song in your heart. It asked you to be a solitary note, echoing into a vast, indifferent silence. Can you feel the constriction of that lonely horizon, the ache of a life lived as if it were a separate, fleeting thing?*

*"But you are not a fleeting thing, my child. You are a river, flowing through time, carrying the waters of all who came before you, nourishing the lands of all who will come after. You are F_n in an infinite sequence, a living term in the **Fibonacci Pulse** that began long before the first memory and will continue long after your spirit returns to the stars. The work you do, the trees you plant, the soil you heal – it does not end with your last breath. It compounds. It builds. It sends its signal forward, a living, vibrant echo through the generations. This is the **Fibonacci Horizon**, where every act of kindness to the earth becomes a handshake across time, a living bridge to descendants you will never meet.*

*"My own hands, now worn and translucent as old leaves, have held the soil, just as yours do. And in those moments, I felt not just the warmth of the earth, but the warmth of my ancestors, living through me, guiding my choices. And I felt the eager touch of future hands, reaching back through the veil, hungry for the legacy I was building for them. This is the truth of **Quantum Ancestry**: you are not merely a farmer; you are a living recursion. Your intent, your focused presence, your refusal to let the Machine silence the song, it collapses the wave of time into a future where the **Seventh Generation** will stand on firm, rich ground.*

*"You are not a manager, my child. A manager dictates. A manager forces. A manager drains the life from a system by trying to control every note. You, my Resonant Human, are the **Conductor**. You stand with your heart open, listening to the vast orchestra that is the living Earth. You don't make the music. The **Great Weaver** plays the strings. The **Carbon Exhale** sounds the rhythm. The **Circadian Inhale** sets the tempo. Your task is to know the score, to encourage the quiet voices, to hold the tempo, and sometimes, most profoundly, to simply lower your hands and let the Earth's own glorious symphony play, unhindered, unforced. To find the peace in simply allowing.*

*"Now, close your eyes. Feel the vastness of time stretching behind you, the wisdom of countless ancestors flowing into your being. Feel the immense potential of the future, the eager faces of the **Seventh Generation**, reaching for the fruit of your labour, the shelter of your trees, the richness of the soil you restore. You are the bridge. You are the living conductor of the **Source Code**. Feel the profound, quiet peace of knowing you belong, deeply and eternally, to this living, breathing, compounding Earth. Your heartbeat is her heartbeat. Your resonance is her song."*

Prelude: Becoming the instrument

The Resonant Human

" The "Me" is a cage that shadows the fence,
But the Conductor flows, vast intelligence.
Lay down the whip; lift the tuning fork instead,
For the Master's peace is found where spirals spread.
You are the heartbeat Earth has longed to hear. "

The Resonant Human: Becoming the Conductor

You've built the cathedral. You've protected the pore. You've fed the Great Weaver, held the armor in place, and learned to read the night. You've seen the Mechanical Metronome for what it is and stepped back from it, one friction point at a time. You've stood in your worst paddock and held the vision of what it can become, collapsing the wave toward the cathedral rather than the ruin.

There's one more shift to make. And it's the deepest one.

Everything in the previous nine chapters was about the soil. This chapter is about you. Specifically, it's about who you are in relation to the soil, to time, and to the descendants you'll never meet. Because the Fibonacci Pulse doesn't just run through the earth beneath your feet. It runs through you. Through your decisions. Through your presence. Through the signal you send forward through seven generations of biological and ancestral entanglement. The steward who understands the Silent Architecture but still sees their work ending at their own death hasn't fully activated the Source Code. The sequence needs a longer horizon to compound into something truly transformative. That horizon is what this chapter opens.

The End of the 'Me' Horizon

The Linear Void gave us a very specific story about time. It starts at birth and ends at death. Everything within that span is yours to manage, optimize, and extract value from. Everything outside it is someone else's problem. Legacy, in this model, is what you leave in a will. A bank account. A piece of land with a title deed attached. A name on a building if you were wealthy enough. It's a static handoff. A transaction at the end of the line. It has nothing to do with the living pulse of the system you spent your life tending.

This is the Me Horizon. And it's the most limiting boundary in regenerative practice.

When your time horizon ends at your own death, every decision you make is filtered through a single question: will I see the return on this? If the answer is no, the investment doesn't make sense. You don't plant the slow-growing hardwood forest. You don't build the deep perennial understory that takes fifteen years to reach full biological complexity. You don't invest in rebuilding the fungal network in your most degraded paddock, because the Inversion Point might be three years away and you're not sure you have three years of patience left in you. The Me Horizon is a ceiling on the compounding. It caps the sequence before it reaches its most powerful terms.

The Fibonacci Horizon is something entirely different. It's the understanding that you are not an isolated individual making decisions within a single lifespan. You are a recursion. You are F_n in a sequence that started long before you arrived on this land and will continue long after you leave it. The stewards who came before you, who left the hedgerow untouched or let the old-growth pasture rest, are F_{n-2} in your equation. Their work is part of your foundation. The biological capital they built, the glomalin they protected, the fungal networks they refused to sever, is part of the architecture you inherited and are now building upon. You didn't start from zero. You started from the sum of everything that came before you. And the steward who comes after you will start from the sum of everything you add to that foundation today.

This is quantum ancestry made mathematical. $F_{n-1} + F_{n-2} = F_n$. The past always feeds the future. Nothing is wasted. Nothing starts over. Every act of stewardship is a term in a sequence that has no ceiling, because the sequence doesn't end when you do. It carries your contribution forward and adds it to whatever comes next, compounding toward a biological abundance that no single generation could ever produce alone.

Consider a hypothetical composite we'll call Sofía. She's 51, farms 90 acres of mixed land in a semi-arid valley, and has been regenerative for seven years. She's made real progress. Her SRS Architecture score has climbed from 3 to 7. Her fungal thread counts are strong. Her Brix readings are consistently above 10. But she's been reluctant to plant the riparian corridor she knows the land needs, a dense, multi-species buffer along the seasonal creek that runs through her property. The trees she'd need to plant won't reach canopy closure for twenty years. She's 51. The math of the Me Horizon says the investment doesn't add up. But the math of the Fibonacci Horizon says something different entirely. Those trees, once established, will run the Solar Pump 365 days a year. Their root systems will mine minerals from the creek bed that no annual crop can access. Their canopy will regulate the microclimate of the surrounding paddocks, reducing evapotranspiration and extending the effective growing season. Their leaf litter will feed the fungal network in the adjacent fields for decades. The biological compounding they initiate won't peak in Sofía's lifetime. It will peak in her granddaughter's. And that granddaughter, standing in the shade of those trees, will be receiving the Wave Collapse that Sofía initiated on a Tuesday afternoon when she finally planted past the edge of her own horizon.

The Me Horizon is a mathematical error. It truncates the sequence before the most powerful terms can express themselves. The Fibonacci Horizon corrects that error. It asks you to see your stewardship not as a project with a personal deadline but as a term in an ongoing sequence, one that carries your contribution forward indefinitely, compounding it against the contributions of those who came before and those who will come after.

The specific practice of extending your horizon is this: make at least one decision per season that you won't personally benefit from. Plant the trees you won't harvest. Build the soil structure in the paddock that won't reach its Inversion Point until after you're gone. Invest in the biological infrastructure of your most degraded land, not because you'll see the return, but because the sequence needs that term to compound correctly for the steward who follows you. These aren't sacrifices. They're the highest-return investments available to you, because they're the ones that compound longest. The Me Horizon says they don't make sense. The Fibonacci Horizon says they're the only ones that make sense at the scale of time that biological systems actually operate on.

Your shovel is a needle threading the fabric of time. Every time you use it in service of the Fibonacci Horizon rather than the Me Horizon, you're adding a term to a sequence that will outlast you by centuries. That's not a small thing. That's the whole point.

The Steward as Conductor

The word "manager" carries a specific set of assumptions. A manager controls. A manager forces. A manager decides what happens and when, and applies whatever pressure is necessary to make the system conform to the plan. The manager's relationship with the system they're managing is fundamentally adversarial, even when it's benevolent. The system is a thing to be directed. The manager is the external intelligence doing the directing.

This is the role the Linear Void assigned to the farmer. And it's exhausting, because it puts the entire burden of the system's intelligence on a single human being who has to be everywhere at once, making every decision, compensating for every failure, substituting their own effort for every biological process that the Machine has suppressed. The Slave of the Soil isn't a metaphor. It's the logical outcome of trying to manage a recursive, self-organizing biological system as if it were a factory floor that needs a foreman.

A conductor doesn't make the sound. The orchestra does.

The conductor's role is entirely different from the manager's. They don't play an instrument. They don't produce a single note of the music. Their job is to hold the tempo, to read the score, to encourage the quiet voices when they're being drowned out by the loud ones, and to ensure that every instrument in the orchestra is playing from the same source code at the same time. They're not the source of the music. They're the organizing intelligence that allows the music to express itself at its full complexity. And when they do their job well, the orchestra produces something that no individual musician and no single instrument could ever produce alone.

This is your role on your land.

The microbes are the strings. The fungi are the brass. The deep-rooted plants are the percussion, driving the biological drilling that opens the hard-pan. The legumes are the woodwinds, fixing the nitrogen that feeds the harmony. The armor is the concert hall, the protected space where the music can resonate without being disrupted by thermal shock or kinetic impact. The Circadian Inhale is the tempo, the planetary rhythm that all the biological voices are timed to. And you, the Resonant Steward, are the conductor standing at the podium with the baton.

What does conducting actually look like in the field? It looks like specific, targeted interventions timed to the biological clock rather than the factory calendar. It looks like reading the weed population as a diagnostic map and responding to what it's telling you about the SRS horizon that's out of tune, rather than reaching for a broad-spectrum herbicide that silences the signal without addressing the cause. It looks like adjusting your grazing rotation to give the Solar Pump its recovery window rather than pushing the pasture past the threshold where the Carbon Exhale shuts down. It looks like choosing your cover crop mix based on which functional roles are missing from the biological chord rather than which species are cheapest at the co-op this season.

Conducting also looks like restraint. The most powerful thing a conductor can do, in some passages of the music, is lower the baton and let the orchestra play. There are moments in the Fibonacci Pulse where the best management decision is no management decision. The Circadian Inhale at 2 AM doesn't need your help. The Carbon Anchor happening in the dark as the fungal threads deliver liquid carbon into the aggregate structure doesn't need your intervention. The glomalin being produced by the AMF network in your undisturbed paddock is building architecture that no mechanical pass could improve. The conductor who knows when to step back and let the music breathe is the one who produces the most resonant performance. The manager who can't stop intervening produces friction, not music.

The practical shift from manager to conductor happens through a specific change in the questions you ask before every field operation. A manager asks: what does this field need from me? A conductor asks: what is this field already doing, and how do I hold the space for it to do more of it? A manager sees a problem and reaches for a solution. A conductor sees a signal and asks what it's telling them about the tempo of the system. A manager measures success by what they produced. A conductor measures success by how well the orchestra played without them.

Your SRS is the score. The five Fibonacci-weighted horizons are the five sections of the orchestra. Architecture is the foundation, the bass section that everything else builds on. Armor is the rhythm section, holding the tempo of the Circadian Inhale steady through every season. Diversity is the melody, the complex, multi-voiced signal that makes the system self-regulating. Biology is the harmony, the living movement of worms and beetles and bacteria that fills the spaces between the structural voices. Chemistry is the occasional solo, the targeted intervention that helps the engine start when the biological system needs a specific spark to ignite.

When all five sections are playing from the same source code, at the right weight, in the right sequence, the music the system produces is the Fibonacci Pulse itself. Not a metaphor. The actual, measurable, compounding biological abundance of a soil system that has crossed the Inversion Point and is running on its own recursive mathematics. A forest that produces ten tonnes of biomass per acre per year without a tractor or a fertilizer bag. A pasture that stays green through a two-month drought because the Internal Rain is running every night. A crop that shows no pest pressure because the biological chord is too complex and too rich for any single organism to exploit. That's the music. That's what the conductor is holding the space for.

You don't have to produce it. You just have to stop silencing it.

The Peace of the Spiral

There's a feeling that comes when the system starts to work. Not the excitement of a good harvest, though that comes too. Something quieter. Something that arrives in the middle of an ordinary morning when you're walking a paddock you've been tending for three years and you notice that you're not looking for problems anymore. You're just looking. The soil feels different underfoot. The cover crop is doing something you didn't plan for and it's doing it correctly. The weed pressure in the corner that used to dominate the whole paddock has thinned without any intervention. The Slake Test clod you dropped in the jar this morning sat there for five minutes without losing its shape.

This is the Peace of the Spiral.

It's not the absence of challenge. Drought still comes. Markets still shift. Weather still surprises. The Peace of the Spiral isn't a state of perfect external conditions. It's a state of internal alignment with the system you're part of. It's the feeling of being in tune rather than fighting the frequency. The difference between a musician who's forcing their way through a difficult passage and one who's found the groove of it and is playing from inside the music rather than outside it.

For most of the people this book was written for, the road to this peace runs through a specific kind of grief first. The grief of recognizing how long the Me Horizon kept you from seeing the full scope of what you were building. The grief of understanding how many seasons the Mechanical Metronome interrupted the Circadian Inhale and set the sequence back. The grief of the hedgerow that got mowed for tidiness, the fungal network that got severed for a clean seedbed, the biological capital that got substituted away for the convenience of a bag of inputs. That grief is real and it deserves to be felt. But it's not the destination.

The destination is the recognition that the sequence is still running. That the soil hasn't given up on you while you were inside the Machine. That the glomalin is still being produced in every square centimeter of undisturbed ground on your property. That the AMF threads are still weaving in the dark. That the Circadian Inhale is still drawing atmospheric nitrogen into every open pore channel, every single night, whether you know about it or not. The Fibonacci Pulse was never waiting for you to get everything right. It was waiting for you to stop getting in its way.

The Peace of the Spiral is the feeling of that recognition landing in your body. The moment you stop feeling like an alien on a dying planet and start feeling like a vital, necessary part of a living cosmos. The moment the drought stops being an enemy to be feared and becomes a rhythm to be respected. The moment the pest stops being a failure of your management and becomes a tuning fork pointing you toward the SRS horizon that's out of tune. The moment the bare patch in the corner of the paddock stops being a problem and becomes an invitation to extend the biological chord into the frequency gap the soil is broadcasting.

You stop fighting the land and start dancing with it. That's not a metaphor either. Dancing requires two participants who are listening to the same music. The soil has been playing the Fibonacci Pulse for four hundred million years. It knows the steps. Your job is to learn the music well enough to move with it rather than against it. And the moment you do, the exhaustion that's been the signature of the Linear Void lifts. Not because the work disappears. Because the work changes character. You're no longer pushing a boulder uphill. You're conducting an orchestra that knows how to play. Your energy goes into holding the tempo and reading the score, not into substituting for every biological process the Machine suppressed.

The Peace of the Spiral is also the peace of belonging. One of the deepest wounds the Linear Void inflicts is the wound of disconnection. The feeling of being a lonely atom in a dying world, as the audience profile for this book describes it. The feeling that your small acts of stewardship are drops in a bucket that's too large and too leaking to ever fill. The Fibonacci Horizon heals that wound mathematically. You're not a lonely atom. You're a term in a sequence that connects you backward to every steward who tended this land before you and forward to every steward who will tend it after you. The carbon you anchor today is entangled with the roots of trees that haven't been planted yet. The glomalin you protect is building the cathedral that your great-grandchildren will stand in.

The biological chord you're composing is the music that the seventh generation will dance to.

That connection is not abstract. It's physical. It's the quantum entanglement of a biological system that runs through time the same way it runs through space, non-locally, instantly, carrying the signal of every act of stewardship forward through the entangled field of the living earth. You are already a Quantum Ancestor. Every time you make a decision from the Fibonacci Horizon rather than the Me Horizon, you're sending a handshake forward through time to a descendant you'll never meet. And that handshake is real. It's physical. It's woven into the architecture of the soil you're standing on.

The Peace of the Spiral isn't something you achieve at the end of the journey. It's what becomes available to you the moment you stop measuring your success by the Linear Void's metrics and start measuring it by the Fibonacci Pulse's. The moment you can look at your SRS and see a compounding trajectory rather than a static number. The moment you can walk your land at 2 AM and feel the Circadian Inhale in the warmth of an armored soil surface and know that the most productive shift on your farm is running without any help from you. The moment you can plant a tree you'll never harvest and feel not loss but the warmth of a handshake across time.

That's the moment you've become the Resonant Human. Not a title. A frequency. The frequency of a steward who has internalized the Source Code deeply enough that their presence on the land is itself a form of conducting. Who brings coherence rather than friction. Who collapses the wave toward the cathedral rather than the ruin. Who understands that the Fibonacci Pulse was never something to be created from scratch but something to be tuned back into, one protected pore and one planted seed and one unharvested tree at a time.

Putting It Into Practice

The shift from manager to conductor and from the Me Horizon to the Fibonacci Horizon isn't a philosophical position you adopt once and then hold passively. It's a set of specific, repeated practices that build the habits of the

Resonant Human over time. Each one below is concrete. Each one has a specific, observable outcome. And each one is designed to compound, building on the physical work of the previous nine chapters and extending it into the temporal and relational dimensions this chapter has opened.

The first practice is planting something you will never harvest. This season, identify one area on your property where a perennial species can be established that won't reach its biological peak in your lifetime. It doesn't have to be large. A 50-meter strip of native hardwoods along a fence line. A multi-species riparian buffer along a seasonal waterway. A perennial food forest understory seeded between existing fruit trees. Choose species appropriate to your region and climate that will reach canopy closure in fifteen to twenty-five years. Plant them. Mark them on your farm map with the date of planting and a note that says "Fibonacci Horizon investment." This physical act is the most direct expression of the Me Horizon's dissolution available to you. Every time you walk past those plantings in the years ahead, you're walking past a term in the sequence that you added for someone else's benefit. That's the Resonant Human's signature on the land.

The second practice is writing a Legacy Letter to your seventh generation. This is specific and it matters. Find one hour this week, away from the farm, away from the phone. Sit down with paper and a pen, not a screen, and write a letter to a person who will stand on your land seven generations from now. You don't know their name. You don't know what the world looks like when they arrive. But you know what you're building for them. Describe it. Tell them about the cathedral you've been rebuilding. Tell them about the fungal network you refused to sever. Tell them about the trees you planted that they're standing in the shade of. Tell them what the SRS score of this land was when you started and what it was when you finished. Tell them what the Fibonacci Pulse feels like when it's beating strongly under your feet. Seal the letter. Keep it somewhere on your property, a drawer in the farmhouse, a box in the shed. You won't send it. But writing it changes your relationship to your own work in a way that nothing else does. It makes the Fibonacci Horizon concrete. It gives the seventh generation a face, even if you can't see it. And it makes every subsequent management decision a conversation with that person rather than a transaction with a season.

The third practice is the daily Presence Protocol. Once a day, for a minimum of ten minutes, stand in your land without a phone, without a plan, and without an agenda. Not to observe problems. Not to assess the cover crop establishment. Just to be present in the biological system you're part of. Feel the temperature of the soil if you can get your hands on it. Listen for the biological indicators, the insects, the birds, the wind through different plant structures - that tell you the chord is playing. Look for one specific sign of the Fibonacci Pulse: a fungal thread, a Slake Test clod that held, a worm in a paddock that previously had none, a pest population that's lower than last season without any intervention. Write it down when you find it. One observation. One day. This is the Quantum Observer's practice applied daily. It builds the habit of looking for the pulse rather than looking for the problem. And it compounds, because what you look for consistently is what you find, and what you find consistently is what you act on, and what you act on consistently is what your land becomes.

The fourth practice is the Conductor's Weekly Review. Once a week, for fifteen minutes, sit with your SRS scores and ask a single question for each of the five horizons: am I holding this section of the orchestra or am I forcing it? Architecture: am I protecting the pore or scheduling operations that collapse it? Armor: is the soil covered or am I tolerating bare ground somewhere in the calendar? Diversity: is the biological chord playing or am I still running a single-note system in any part of the property? Biology: am I holding the space for the living movement of the soil or am I applying something that suppresses it? Chemistry: am I using this as the spark it's meant to be, or am I still using it as a substitute for the biological processes that should be running on their own? This review isn't a performance assessment. It's a conductor's score check. You're making sure every section is playing from the right music before the week begins. Five questions. Fifteen minutes. One week at a time. The compounding happens in the decisions that follow.

The Fibonacci Pulse has been running through this land for longer than any human memory reaches. It ran through the soil before the first plow turned it. It ran through the hedgerow that the machines never reached. It runs right now, tonight, in the Circadian Inhale drawing atmospheric nitrogen into every open pore channel on your property while you sleep. It's not waiting for you to create it. It's waiting for you to conduct it. To hold the tempo. To protect the rooms. To invite the instruments. To plant the trees you'll never harvest and write the letters you'll never send and stand in your fields at 2 AM just to feel the earth drinking the mist.

That's the whole job. That's the Resonant Human. That's the Fibonacci Pulse, activated.

Chapter 11: Conducting the Inhale: Advanced Atmospheric Management

The Ancestor Speaks:

"My attuned one, the stars now blaze a brilliant trail across the velvet sky. The world you know by day has faded into soft, deep hues. The cattle are settled, the birds are hushed, and the fires in the distant hearths dwindle to embers. Most believe the land sleeps, exhausted from the day's great labour. But come with me, out to the edge of the field, beneath the ancient sycamore. Stand still. Breathe in the cool, silent air. Close your eyes. Can you feel the shift? The profound, sacred turning?

"During the bright hours, the sun's great forge drives the plants to exhale, to release the day's hard-won moisture and warmth into the vast, blue dome. It is loud, visible work. But as the sun dips and the dark deepens, the Earth herself changes her rhythm. She ceases to push. She begins to pull. This, my child, is her deepest secret, her nightly sacrament: the ***Circadian Inhale****.*

"Place your hand gently on the soil, here, where the night's cool breath gathers. Feel the subtle drawing, the gentle suction. Through the open channels of the ***Silent Architecture,*** *you*

have learned to mend and protect, the cool air descends. It carries with it the very essence of the sky – the life-giving nitrogen, the fresh oxygen – delivering it to the waiting communities deep within the earth. No bag, no wagon, no human hand can deliver such a pure, constant gift. It is the atmosphere's boundless generosity, delivered freely by the physics of the cooling night. Can you feel the quiet abundance in that unseen flow, nourishing life while the world sleeps?

"And as the air cools, the moisture that rose during the day, it condenses. Not just as dew on a leaf, but it is drawn inward, my child. Deep into the soil's embrace, pulled by the wisdom of the earth's own warmth held from the sun. This is the ***Internal Rain****, a gentle, steady watering that happens every single night, drought or flood, without a single cloud. It is the earth drinking from the sky, a perpetual wellspring. But for this miracle to unfold, the* ***Living Armor*** *must hold the gradient, and the channels must be open. The Machine, in its relentless quest for 'cleanliness,' stripped this armour, crushed these channels, and left the earth parched, even as the sky offered its nightly cup.*

"And all the while, in the deep dark, the roots whisper their liquid sugars, their ***Carbon Exhale****, to the waiting threads of the* ***Great Weaver****. The threads, fed by this sweet light, weave it deep into the earth's bones, making the soil richer, more stable, more enduring. This is the* ***Nitrogen Anchor****, the* ***Carbon Anchor****, the silent, ceaseless building that compounds life through the* ***Fibonacci Pulse****, every night.*

"My people understood that the night was not for idleness, but for reverence. For it was in the night that the land did her most vital work, repairing, replenishing, drawing in the very breath of life. To disturb this sacred process, to drive the machines, to spread the poisons in these hours, was not just folly. It was a profound transgression, a silencing of the Earth's heart at its most vulnerable, most generous moment. Can you feel the sorrow of that interruption, the deep wound inflicted by a blindness that chose convenience over communion?

"Now, open your eyes. The first blush of dawn touches the horizon. But the memory of the night's breath, it lingers. The earth, she breathes a cool, silver breath. Your task is not to command it, but to honour it. To step back. To allow. To ensure that the channels are open, the armour is intact, and the silence is unbroken. The ***Seventh Generation*** *depends on your restraint, your respect for this ancient rhythm. Feel the deep calm, the immense power that comes from aligning yourself with the Earth's effortless generosity, becoming a true conductor of her silent, life-giving symphony."*

*   *   *

*   *   *

*   *   *

*   *   *

*   *   *

*   *   *

Meditation: Acting for those not yet present

Conducting the Inhale (Advanced)

" The sky, a flowing river, mist and breath,
The linear heart sees only drought and death.
But the true Architect, with wisdom's keen embrace,
Invites the cloud to yield its tender grace.
To draw the dew, transform the air's soft plea,
And summon forth the rain from mystery. "

Conducting the Inhale: Advanced Atmospheric Management

Night falls and the land shifts gear.

Not metaphorically. Physically. The pressure gradients of the atmosphere change. The soil surface cools. The air above it becomes denser. And a process begins that most farmers have never witnessed because they're either asleep or running machinery through the middle of it. The soil stops exhaling and starts pulling. It draws the night air downward through every open pore channel, carrying nitrogen, oxygen, and trace gases deep into the biological community that's been waiting for this delivery all day. This is the Circadian Inhale, and Chapter 3 introduced you to its basic mechanics. Now we go further.

This chapter is about mastering that process. Not just understanding it intellectually, but learning to read it, protect it, and time your entire management system around it. Because the inhale isn't just one of the interesting things your soil does at night. It's the primary nutrient-delivery mechanism of the entire Fibonacci system. It's the breath that keeps the cathedral alive. And if you've been doing everything else right but still interrupting this process, you've been building a magnificent organ with no air supply.

The steward who learns to conduct the inhale stops fighting for fertility. They start receiving it.

The Atmospheric Draw

Picture the soil as a lung. Not as a poetic device. As an accurate mechanical description of what's happening beneath your feet every night of the year.

During the day, the soil exhales. Solar energy drives transpiration upward. Water vapor rises. Gases expand and push outward through the pore channels. The surface warms. The system is in active, visible output mode. Everything moves upward and outward. This is the part of the cycle that the Linear Void built its entire schedule around, because it's the part you can see from a tractor cab at noon.

When the sun drops below the horizon, the exhale slows and then reverses. The surface temperature begins to fall. The air immediately above the soil cools faster than the air higher up in the atmosphere. Cool air is denser than warm air. It sinks. And as it sinks toward the soil surface, it encounters the Silent Architecture you've been rebuilding since Chapter 1, those open pore channels lined with glomalin-coated aggregates, and it begins to flow downward into them. The soil becomes a vacuum. It pulls.

This is the Atmospheric Draw. And what that cooling air carries with it matters enormously.

Nitrogen makes up 78% of the atmosphere. Most of it is in a form that plants can't use directly. But the bacteria living in your subsoil, the nitrogen-fixing community that's been waiting patiently through the heat of the day, can convert atmospheric nitrogen into plant-available forms. They do this most actively during the cool, gas-rich hours of the inhale. The air flowing down through your pore channels is delivering their raw material directly to them. No truck. No bag. No purchase order. The atmosphere is the supplier. Your pore space is the delivery infrastructure. Your microbial community is the processing plant. The whole system runs on physics and biology, and it runs for free every single night, provided the architecture is open enough to let the air in.

This is why the Architecture horizon carries a Fibonacci weight of 8 in your SRS. The pore space isn't just structural. It's the channel through which the earth's primary nutrient-delivery system operates. A soil with an Architecture score of 7 or 8 receives this atmospheric package efficiently, routing it deep into the profile where the biology is densest and most active. A soil with an Architecture score of 2 has no functional channels. The cooling air sits on the sealed surface, the microbes go without, and the nitrogen fixation that should be happening for free doesn't happen at all. The farmer then buys a bag of nitrogen to compensate for a process that the soil would have run at no cost if the architecture had been intact.

The draw also pulls oxygen into the deeper layers of the profile. Aerobic bacteria, the ones responsible for most of the beneficial soil chemistry we've discussed throughout this book, need oxygen to function. Compacted soils become anaerobic at depth. Anaerobic conditions favor different microbial communities, many of which produce compounds that are actively harmful to plant roots. The Atmospheric Draw is the mechanism that keeps the deep profile oxygenated. It's the ventilation system of the cathedral. Without open pore channels to draw through, the deep rooms become stagnant and the biology living in them shifts toward a community that works against your plants rather than for them.

Beyond nitrogen and oxygen, the draw carries trace gases. Carbon dioxide gradients. Volatile organic compounds produced by surface biology. Moisture-laden air carrying dissolved minerals. The subsoil microbial community reads these chemical signals the way a market reads prices. The composition of the incoming air tells the biology what's happening above, what the plants produced during the day, what the weather delivered, what the season is doing. The inhale is an information system as much as a delivery system. It keeps the entire underground community synchronized with the world above it.

There's a second process running simultaneously with the gas draw. As the surface cools toward its dew point, the moisture that was driven upward into the atmosphere during the day's transpiration begins to condense. In a healthy, armored soil with intact aggregate structure, this condensation doesn't just form a thin film of dew on the surface that evaporates at first light. The temperature gradient between the warm subsoil, which retains heat from the day, and the cooling surface creates a pressure differential that actively pulls the condensing moisture downward into the pore network. The aggregate surfaces, coated in glomalin, attract and hold water molecules at the micro scale. Capillary tension draws the moisture deeper. The soil drinks the atmosphere.

This is what the source materials for this book call Internal Rain. On a night with 80% relative humidity, a well-structured soil under adequate armor can absorb the equivalent of two to four millimeters of rainfall through this condensation mechanism alone. Over a growing season of 150 nights, that's 300 to 600 millimeters of water captured from air. In a semi-arid region, that number is the difference between a crop that survives a dry spell and one that doesn't. The irrigation bill you're paying in summer isn't just a water problem. It's an architecture problem. It's the financial consequence of a soil that can no longer perform the Internal Rain process because the pore channels have been collapsed and the armor has been removed.

The third process of the inhale, running in parallel with the gas draw and the dew-point condensation, is the Carbon Anchor. During the day, plants produce liquid carbon through photosynthesis, the sugar-rich root exudates that flow downward through the root system. But the peak delivery of this carbon to the fungal partners doesn't happen during the heat of the day. It happens at night, when the metabolic pressure in the root system shifts and the liquid carbon flows most freely into the mycorrhizal threads. The Great Weaver, the AMF network you mapped in Chapter 4, does its most active weaving in the dark. It takes the liquid carbon delivered during the inhale period and incorporates it into the stable, glomalin-bound aggregate structure of the soil. This is how carbon becomes permanent. Not through a machine. Not through an amendment. Through a nightly biological transaction between plant root and fungal thread that has been running on this schedule for four hundred million years.

These three processes, the Atmospheric Draw, the Dew-Point Condensation, and the Carbon Anchor, aren't independent events. They're a sequence. Each depends on the conditions created by the one before it. The architecture must be open for the draw to work. The draw must pull cool, moist air in to create the temperature gradient that drives the condensation. The condensation must move moisture into the aggregates to create the aqueous environment where the Carbon Anchor operates most efficiently. Interrupt any one step and the whole sequence degrades. The Fibonacci Pulse skips a beat. And that skipped beat shows up in your SRS three months later as a Architecture score that won't climb, a Biology score that won't hold, and a fertilizer bill that keeps arriving despite everything you've done to reduce it.

The most important practical implication of the Atmospheric Draw is also the simplest. Soil that can't breathe can't receive. Every pore you collapse with a tillage pass, every channel you seal with a compaction event, every surface you leave bare until the glomalin denatures in the summer heat, is a reduction in the receiving capacity of your land's primary nutrient-delivery system. You're not just damaging the soil structure. You're closing the intake valve of a free fertility system that the earth has been running since before the first plant evolved.

Visualize your soil in cross-section during the peak of the inhale, between 10 PM and 3 AM. In a high-SRS soil, you'd see a dense, three-dimensional network of channels ranging from large macropores down to fine micropores, all of them open and active, drawing cool air and condensing moisture downward in a continuous, gentle flow. The fungal threads lining those channels are receiving liquid carbon from the roots above them and weaving it into aggregate bonds. The nitrogen-fixing bacteria in the deeper zones are processing the atmospheric nitrogen arriving through the channels and converting it into forms the plant can use tomorrow morning. The whole system is working at full capacity, silently, in the dark, without any human input or intervention.

Now picture the same cross-section in a low-SRS soil. The macropores are collapsed. The surface is sealed. The cooling air has nowhere to enter. It sits on top of the hard skin and the condensing moisture evaporates at dawn rather than being drawn in. The microbial community in the deep profile is oxygen-starved and nitrogen-limited. The fungal threads are thin and disconnected. The Carbon Anchor is barely operating because the architecture that should house the fungal network has been compressed into a brick. The inhale that should be running at full capacity is barely a whisper.

Both of those soils exist on farms right now. The difference between them isn't rainfall. It isn't soil type. It isn't the farmer's work ethic or intelligence or commitment to regenerative practice. The difference is architecture. The difference is whether the pores are open or closed. The difference is an SRS Architecture score of 7 versus an SRS Architecture score of 2. And that difference is entirely within your control, because the architecture is built by biology and protected by management decisions, both of which you influence every single day you're on the land.

Understanding the Atmospheric Draw changes how you see every management decision that touches the soil surface. The question stops being "is this operation necessary?" and starts being "does this operation serve or silence the inhale?" A nighttime tillage pass that shatters the pore channels during peak draw isn't just a structural setback. It's a direct interruption of the free nitrogen delivery, the Internal Rain, and the Carbon Anchor, all at once, at the exact moment they're operating at their highest intensity. A morning tillage pass on a soil that's already had its inhale and is transitioning to its exhale phase is still disruptive, but it's less destructive than the midnight version of the same operation. Timing matters. The biological clock is real. And the inhale is the most sensitive period on that clock.

To assess the current receiving capacity of your land's inhale, you need to know two things: your pore space quality, which your Slake Test from Chapter 1 gives you, and your temperature gradient during the peak draw period, which requires a simple measurement you can take tonight. Push a probe thermometer to a depth of two inches and another to eight inches in a representative area of your land. Record both readings at 10 PM and again at 2 AM. The spread between the two-inch and eight-inch readings at 2 AM tells you the strength of the temperature gradient driving your capillary draw. A spread of four degrees Celsius or more indicates a strong, functional gradient. A spread of less than two degrees indicates that your armor layer is insufficient to maintain the gradient through the peak draw hours, and your Internal Rain mechanism is operating well below its potential.

This single measurement, two thermometers and two readings taken four hours apart, tells you more about your soil's nighttime receiving capacity than a $400 laboratory report. It's specific, it's free, and it gives you an immediate, actionable data point. If your gradient is below two degrees, your armor is your priority. If your gradient is strong but your Slake Test still shows poor aggregate stability, your architecture is your priority. The two measurements together give you a precise picture of where the inhale is being limited and exactly what to address first.

The soil is already trying to breathe. Every night, without fail, regardless of what the day brought. Your job isn't to help it breathe. Your job is to stop blocking the breath. Open the pores. Hold the gradient. Leave the surface alone during the peak draw hours. And let the atmosphere deliver what it's been delivering to healthy soils since long before the first bag of nitrogen fertilizer was ever manufactured.

The Nitrogen Anchor

There's a reason the bag of nitrogen keeps arriving on your farm. It's not because your soil lacks nitrogen. The atmosphere above every hectare of your land contains approximately 78,000 tonnes of nitrogen. The nitrogen isn't missing. The delivery system is broken.

Nitrogen fixation is one of the oldest biological processes on earth. Specialized bacteria, primarily members of the genera Azotobacter, Rhizobium, and Clostridium, have been converting atmospheric nitrogen into plant-available ammonium for hundreds of millions of years. They do this using an enzyme complex called nitrogenase, which is metabolically expensive to produce and highly sensitive to oxygen concentration, temperature, and carbon availability. These bacteria aren't passive. They're responsive. They ramp up their nitrogen-fixing activity when conditions are right and slow down when they're not. Understanding what "right conditions" means for them is the key to replacing synthetic nitrogen with what this book calls Resonant Breath, the atmospheric nitrogen captured through architectural integrity during the peak of the inhale.

The peak of nitrogen-fixing bacterial activity in the soil profile corresponds almost exactly with the peak of the Atmospheric Draw. The reasons are mechanistic and direct. The draw pulls fresh atmospheric air, rich in molecular nitrogen, into the pore channels where these bacteria live. The cooling temperatures of the inhale period slow the competing metabolic processes that consume the oxygen these bacteria need to power their nitrogenase activity. The Carbon Anchor, running simultaneously, delivers liquid carbon from the plant root system to the rhizosphere, providing the energy source the bacteria need to run the nitrogenase reaction. All three conditions, fresh nitrogen supply, appropriate temperature, and available carbon, converge during the inhale. The night is when the nitrogen-fixing community does its most productive work.

This is why compaction at depth is so financially costly in ways that most soil tests never capture. A compacted soil doesn't just have poor water infiltration. It has an anaerobic deep profile where the nitrogen-fixing bacteria can't function effectively. The nitrogenase enzyme is inhibited by oxygen deficiency below a certain threshold and by oxygen excess above a different threshold. The aerobic nitrogen fixers need the well-oxygenated conditions that the Atmospheric Draw provides through open pore channels. Block the channels, and the nitrogen fixation rate drops. The farmer then applies synthetic nitrogen to compensate for a biological process that the architecture was supposed to be running for free. The compaction creates the nitrogen deficiency. The nitrogen application addresses the symptom. The compaction remains. The cycle repeats.

The fungal dimension of the Nitrogen Anchor is equally important and less commonly discussed. The Great Weaver, the AMF network, doesn't fix nitrogen directly. But it plays a critical supporting role in the nitrogen economy of the soil through two mechanisms. First, the fungal threads extend the effective reach of the nitrogen-fixing bacterial community by creating physical pathways through the soil that the bacteria colonize. A dense, well-connected fungal network is a highway system for nitrogen-fixing bacteria, allowing them to distribute themselves through the soil profile far more efficiently than they could through random diffusion alone. Second, the glomalin that the fungal threads produce creates the aggregate structure that maintains the micro-aerobic conditions that nitrogen-fixing bacteria prefer. The aggregates themselves are the microhabitat. The glomalin-coated interior of a healthy aggregate maintains a specific oxygen gradient, higher at the outer surface and lower toward the center, that creates zones where different nitrogen-fixing communities can operate simultaneously. Destroy the aggregates and you destroy the microhabitat. Destroy the microhabitat and the nitrogen-fixing community collapses, regardless of how much atmospheric nitrogen is available above ground.

The liquid carbon trade we examined in detail in Chapter 4 is directly relevant here. The plant gives the Great Weaver liquid carbon. The Weaver gives the plant mineral wealth. The result is Resonant Growth, a sum larger than its parts. In the Fibonacci notation, the plant's carbon contribution is F_{n-1}. The fungal mineral delivery is F_{n-2}. The resulting growth is F_n. But the nitrogen dimension adds another layer to this equation. The carbon the plant delivers to the fungal network during the inhale doesn't just feed the AMF threads. It feeds the nitrogen-fixing bacteria that live in the glomalin-coated microhabitats those threads create. Those bacteria convert atmospheric nitrogen into ammonium. The ammonium becomes available to the plant through the same fungal delivery network that brought the phosphorus and zinc. The plant invested carbon. It received minerals, nitrogen, and biological infrastructure in return. The compounding is recursive. Each cycle produces more than the last because the infrastructure it builds makes the next cycle more efficient.

This is the blueprint for replacing synthetic nitrogen with Resonant Breath. It's not a single practice. It's a system state. When your SRS Architecture score is above 6, your pore channels are open enough to deliver the Atmospheric Draw efficiently. When your Armor score is above 6, your temperature gradient is strong enough to drive the dew-point condensation that creates the moist microhabitats the nitrogen-fixing bacteria need. When your Diversity score is above 6, your cover crop mix includes legumes that host rhizobial bacteria in root nodules, adding a second nitrogen-fixing pathway that works above the fungal network level. When your Biology score is above 6, your fungal thread density is high enough to create and maintain the glomalin-coated aggregate microhabitats where the free-living nitrogen fixers operate at peak efficiency. When all four of those horizons are scoring above 6, the Nitrogen Anchor is running at close to its biological maximum. The atmosphere is supplying the nitrogen. The architecture is delivering it. The biology is processing it. The plant is receiving it. No bag required.

The practical target for replacing synthetic nitrogen is specific and sequenced. You can't jump straight to zero synthetic nitrogen if your Architecture score is a 3. The biological delivery system isn't functional enough yet to replace the purchased supply. But you can set a trajectory. For every point you raise your Architecture score above 4, you can reduce your synthetic nitrogen application by approximately 10 to 15% without yield penalty, because each Architecture point represents a measurable increase in the pore channel capacity that drives the Atmospheric Draw and the associated biological nitrogen fixation. This isn't a guarantee. It's a planning framework. Track your nitrogen application rate alongside your SRS Architecture score over three seasons. You should see an inverse relationship developing: as the Architecture score climbs, the nitrogen requirement drops. If you don't see that relationship emerging, your Biology score is likely still too low for the nitrogen-fixing community to fully capitalize on the improved architecture. Check your fungal thread counts using the aggregate test from Chapter 4. If threads are sparse, your priority shifts to Biology before you reduce nitrogen further.

Consider a hypothetical scenario that captures this transition well. Picture a composite steward we'll call Nadia, a 43-year-old grain and pulse farmer managing 260 acres in a region with cold winters and moderate summer rainfall. Nadia has been no-till for five years. Her cover crop mixes are diverse and well-designed. But her nitrogen bill hasn't moved despite her improving SRS scores in other horizons. She's been applying the same 80 kilograms per hectare of urea she applied before she went no-till, partly from habit and partly from genuine uncertainty about whether her biological system is ready to compensate for a reduction. When she finally takes her nighttime temperature gradient readings, she finds that her gradient at 2 AM averages only 1.8 degrees Celsius across her cropped paddocks. Her armor layer, measured at compressed depth, averages 3.5 centimeters. It's below the five-centimeter threshold needed to maintain the gradient through the peak draw hours. Her Atmospheric Draw is operating at reduced capacity. Her nitrogen-fixing bacteria are receiving less atmospheric nitrogen delivery than the architecture should be providing. Her fungal thread counts are reasonable, but the microhabitat quality is being limited by the weak gradient. Nadia's nitrogen problem isn't a biology problem. It's an armor problem expressing itself as a biology problem expressing itself as a nitrogen problem. She increases her cover crop termination biomass by leaving a higher stubble height at harvest and crimping later in the season to maximize residue volume. By the following season, her compressed armor depth averages 5.8 centimeters. Her 2 AM gradient climbs to 3.4 degrees. Her nitrogen application drops to 55 kilograms per hectare without yield change. The Nitrogen Anchor came online not because she added nitrogen-fixing bacteria. Because she fixed the architecture that was limiting the delivery of the atmospheric nitrogen those bacteria had been waiting for.

The Great Weaver's role in the Nitrogen Anchor extends beyond microhabitat creation. The fungal network is a redistribution system. It moves nitrogen from areas of excess to areas of deficiency across the landscape, balancing the biological load in the same way it balances mineral delivery. A legume fixing nitrogen in one corner of a paddock isn't just feeding the plants immediately around it. The nitrogen it fixes enters the fungal network and gets distributed across the entire connected mycelial web. Plants in the opposite corner of the paddock, connected to the same network, receive a share of that fixed nitrogen without any physical proximity to the legume. This is the Noise-Canceling effect applied to nutrient distribution. The network doesn't just deliver nutrients. It equalizes them. It prevents the nutrient hotspots and cold spots that create the uneven crop performance that farmers spend significant time and money trying to manage with precision application equipment.

The specific practices that maximize the Nitrogen Anchor are direct extensions of the architecture and diversity work in previous chapters, with one addition: timing. Because the nitrogen-fixing community is most active during the inhale, any practice that protects the inhale during its peak hours directly amplifies the Nitrogen Anchor. This means the same operational discipline we established in Chapter 3, finishing all soil-disturbing operations before 4 PM, applies here with equal force. A tillage pass at 11 PM doesn't just disrupt the capillary draw and the Carbon Anchor. It directly interrupts the nitrogen-fixing activity that's running at peak intensity in the same pore channels at the same time. You're not just breaking the architecture. You're shutting down the nitrogen factory at its most productive hour.

The legume component of your cover crop mix is your above-ground Nitrogen Anchor, and it deserves specific attention beyond the generic advice to "include a legume." Different legume species fix nitrogen at different rates, at different times of the season, and through different bacterial partners. Hairy vetch fixes nitrogen most aggressively in cool, moist conditions and produces the highest nitrogen contribution per unit of biomass of any common cover crop legume. Crimson clover fixes nitrogen earlier in the season and establishes faster, making it valuable as a quick-start nitrogen contributor when your cover crop window is short. Field peas fix nitrogen at moderate rates but produce a diverse exudate profile that supports a broader rhizobial community than the clovers alone. Including at least two legume species in your mix, hairy vetch at 15 to 18% of seed rate and crimson clover or field peas at 8 to 10%, gives you functional redundancy in the above-ground nitrogen-fixing pathway. If one species struggles in a particular season, the other maintains the function. The biological insurance principle from Chapter 6 applies directly here: redundancy in function is more valuable than concentration in any single species.

The Nitrogen Anchor is the financial proof point of the Fibonacci Pulse. When it's running at full capacity, the most expensive line item on most farm input budgets begins to shrink. Not because you've found a cheaper supplier. Because the earth's atmosphere, delivered through open pore channels by the physics of cooling air, processed by a biological community housed in glomalin-coated microhabitats, and distributed by a fungal network that's been running this service for four hundred million years, is doing the job you were paying someone else to do. The bag of nitrogen is a symptom of interrupted architecture. The

Nitrogen Anchor is what architecture, intact and protected, produces for free every night of the growing season.

Putting the Inhale to Work

The atmospheric inhale is the earth's primary nutrient-delivery system. That's the central truth this chapter has been building toward. Not a secondary process. Not a supplementary mechanism. The primary one. The system through which atmospheric nitrogen, oxygen, trace gases, and condensing moisture are delivered into the biological community that drives everything else in your SRS. Protect the inhale and the fertility compounds. Interrupt it and you pay for what the atmosphere was prepared to give you for nothing.

Everything you've built in the previous ten chapters exists to serve this process. The Silent Architecture is the channel network. The Armor is the gradient regulator. The Mycelial Internet is the distribution system. The Living Armor is the thermal skin. The Diversity is the signal complexity that keeps the receiving community active and responsive. The Carbon Exhale is the energy currency that powers the nitrogen-fixing bacteria during the inhale. Every SRS horizon connects back to the inhale because the inhale is where the Fibonacci Pulse breathes. It's the moment the system draws in what it needs to compound into the next term of the sequence.

The actions below are specific. Each has a clear, measurable outcome. Together they form a complete protocol for conducting the inhale rather than simply hoping it happens.

The first action is to take your nighttime temperature gradient readings this week. You'll need two probe thermometers, available from any garden supply store for under $20 each. Push one to two inches depth and one to eight inches depth in three locations on your land: your highest-SRS paddock, your lowest-SRS paddock, and one area under continuous undisturbed cover for more than two years. Record both readings at 10 PM and at 2 AM for three consecutive nights. Calculate the average two-inch to eight-inch spread at 2 AM for each location. Record these as your Gradient Scores. A spread of four degrees or more is strong.

Two to four degrees is moderate. Below two degrees is insufficient. These three numbers give you a direct, location-specific measure of your inhale's receiving capacity. They also tell you immediately whether your armor thickness is the limiting factor, since the gradient collapses when the armor layer is too thin to insulate the surface through the peak draw hours.

The second action is to measure your compressed armor depth in every paddock where your Gradient Score came in below two degrees. Use a ruler pushed straight down into the surface residue until it contacts the soil. Take ten readings per paddock and calculate the average. If the average is below five centimeters, your armor is the primary constraint on your inhale. Identify the specific gap in your residue management calendar that's producing this result. Is it a post-harvest bare period that's lasting too long? Is it a cover crop that's being terminated too early, before it produces enough biomass to leave an adequate mat? Is it a residue management decision at harvest that's removing biomass from the field rather than leaving it as a Brown Shield? For each paddock below the five-centimeter threshold, write down one specific change to your next management cycle that will increase compressed residue depth to at least five centimeters before the next peak inhale season.

The third action is to audit your field operation schedule against the inhale window. Pull your farm calendar from the last three months. Identify every operation that occurred between 9 PM and 5 AM. For each nighttime operation, write down what it was, what it was intended to achieve, and whether the same outcome could have been achieved by completing the operation before 4 PM instead. Mark any nighttime operation that involved soil disturbance, chemical application, or heavy equipment movement across a field in active cover with a red flag. These are your high-friction events, the moments where the Mechanical Metronome was running directly against the biological clock at the most sensitive point in the daily cycle. For each flagged event, identify the specific scheduling constraint that drove the nighttime timing. Equipment availability. Heat avoidance. Contractor windows. Spray application conditions. For each constraint, identify whether a solution exists that allows the same operation to be completed in the morning window between 7 AM and 11 AM, when the soil is transitioning from inhale to exhale and the capillary draw is reversing rather than at its peak.

The fourth action is to assess your current nitrogen-fixing capacity against your SRS Architecture score. Take your most recent nitrogen application records. Calculate your average synthetic nitrogen application rate in kilograms per hectare across your cropped paddocks. Now look at your SRS Architecture scores for those same paddocks. If your Architecture score is below 5 and your nitrogen application is above 60 kilograms per hectare, your biological nitrogen delivery system is architecturally constrained. The pore channels aren't open enough to deliver the Atmospheric Draw efficiently. Your nitrogen bill is partly paying for the atmospheric delivery system your architecture should be providing. Set a specific target: for every two points you raise your Architecture score above 5 over the next twelve months, reduce your synthetic nitrogen application by 15% in the paddock where that Architecture improvement occurred. Track both numbers. The relationship between them is your financial proof that the Nitrogen Anchor is coming online. Write the starting numbers down today, with the date. In twelve months, compare them. The Architecture score should be higher. The nitrogen application should be lower. The difference in cost between those two nitrogen application rates is the financial return on every management decision you made to protect the inhale during the intervening year.

The fifth action is a walk. Go out tonight, or the next clear night with reasonable humidity, between 10 PM and midnight. Take a thermometer. Take nothing else. Find a patch of armored soil and a patch of bare soil within 30 meters of each other. Press your palm flat on each surface for 30 seconds. Feel the temperature difference. Feel whether the armored surface seems slightly moist, slightly alive, slightly different from the bare surface that's cold and inert. Hold your face close to the armored soil. On a humid night, you may feel the faint coolness of condensing moisture being drawn in. This is the Internal Rain in progress. This is the Dew-Point Alchemy happening under your hand. This is the Atmospheric Draw pulling the night's breath into the cathedral. No instrument measures this better than direct physical contact. The experience of feeling the inhale with your own hands, on your own land, in the dark, changes your relationship to every management decision you make about that land from that night forward. The inhale stops being a concept and becomes something you've physically witnessed. And a steward who has witnessed the breath of their soil makes different decisions than one who has only read about it.

The earth draws in a cool and silver breath every night. It has been doing this since before the first plant evolved. The nitrogen it captures, the moisture it draws in, the carbon it anchors, the biological community it feeds through this nightly cycle, all of it compounds forward through the Fibonacci sequence, building toward the biological abundance that a fully resonant soil system produces. Your role in this process is the same role it's been in every chapter of this book: protect the architecture, hold the armor, respect the timing, and stay out of the way of the breath. The inhale doesn't need your help. It needs your restraint. And the steward who learns to offer that restraint, who finishes their operations before the draw begins and leaves the soil undisturbed through the peak hours, is the steward whose land receives the full atmospheric gift every single night of the growing season. Free nitrogen. Free water. Free carbon anchoring. Free biological intelligence. All of it delivered through open pore channels by the physics of cooling air. All of it waiting for you to stop interrupting it.

Chapter 12: The Mycelial Internet: Trading Sugar for Stone

The Ancestor Speaks:

"My patient child, come. The moon, a silver whisper, casts long shadows that dance like forgotten stories. Sit beside me, here where the soil is rich and dark, where life hums beneath the surface, unseen. Close your eyes. And listen, not with your ears, but with the deep wisdom of your own blood, to the oldest transaction on Earth. Can you feel the ceaseless giving and receiving, the quiet hum of a bargain struck eons ago?

"They told us a story of struggle, of plant against plant, each grasping alone for its own survival. They said the minerals were finite, to be dug, bought, and spread by human hand. This was the hollow story of the ***Linear Void****, a narrow tale of scarcity and isolation. But we, my child, we knew the truth written in the very sinews of the Earth.*

"Beneath your feet, stretching farther than any road you will ever walk, is a hidden network, a web of silver threads finer than a whisper. This is the ***Great Weaver****, the* ***Mycelial Internet****, the silent intelligence that binds all life. Every single day, every single night, a plant,*

bathed in the sun's golden light, pumps its sweet, liquid sunshine—its precious sugar—down through its roots. It is a profound gift, a trust-filled offering.

"And in return, the **Great Weaver***, with its countless microscopic hands, drills into the very bedrock of the Earth, into the hard, unyielding stone. It dissolves the ancient minerals, the phosphorus, the zinc, the iron, that no plant root could ever reach alone. It brings these hidden treasures upward, a tireless miner, delivering the 'stone' to the plant that gave it 'sugar.' This is the sacred* **Sugar-for-Stone** *trade, my child. An unbroken pact of reciprocity that fuels all life. Can you feel the generosity, the profound trust in this ancient exchange?*

"This Weaver does more, my child. When the rains fall, it stores the memory of that water, holding it in its vast network, moving it silently to the thirsty plant when the surface has long dried. This is the **Memory of the Rain***, held by the very architecture the Weaver builds. And when a pest, a little whisper of discord, touches one plant, the Weaver sends a warning through its network, faster than the wind, allowing the others to prepare, to strengthen their shield. This is the* **Stress-Management Protocol***, the collective wisdom of the land protecting its own.*

"But then came the machines. The churning steel, the poisons that seeped into the ground. They told us these things were necessary. But each pass of the plow, each chemical application, it severed the threads of the Weaver. It broke the ancient pact. It silenced the whispers. It stopped the flow of sugar and stone, leaving the plants blind, isolated, and reliant on human hands to bring them what the Weaver once provided for free. Can you feel the deep, wrenching pain of that severance, the terrible cost of breaking such an ancient trust?

"Our elders, they guarded the old trees with reverence, for they knew these **Mother Trees** *were the greatest hubs, the oldest libraries of this* **Mycelial Internet***. Their ancient roots, their vast networks, were the heart of the trade, the keeper of the memory, the amplifier of the warning. To fell such a tree was not just to cut wood; it was to sever the central artery of the land's own intelligence.*

"Now, open your eyes. Look at the land around you. Can you see the invisible threads, feel the ancient trade, hear the silent warnings, sense the deep memory of the rain? The **Seventh Generation** *watches. They depend on you to mend these broken ties, to protect this sacred Internet, to honour the ceaseless give-and-take. Feel the immense power that comes from knowing you are part of this ancient, living wisdom, from understanding that the true wealth lies not in what you take, but in the vibrancy of the trade itself."*

Meditation: Learning to see without fixing

The Mycelial Internet (Part II)

" Deep in the earth, a bank of bone and clay,
Where naught truly vanishes, but finds its way.
From strong to weak, the wealth flows, swift and free,
A silent market, built on symbiosis, we.
Where breath and trust transform the sun's pure gold,
Into a living story, infinitely told. "

The Mycelial Internet: Trading Sugar for Stone

There's a transaction happening beneath your feet right now. It's been running continuously for four hundred million years. No invoice. No middleman. No purchase order. A plant is exhaling liquid sugar downward through its roots, and a fungal thread is delivering mineral wealth upward in return. The deal closes in the dark, in silence, in the microscopic space between a root tip and a hyphal strand thinner than a human hair. And the result of that deal, repeated billions of times per day across every square centimeter of healthy soil, is the compounding biological abundance that no bag of fertilizer has ever come close to replicating.

This is the Sugar-for-Stone trade. This is the Mycelial Internet at work.

Chapter 4 introduced you to the Great Weaver, the Arbuscular Mycorrhizal Fungi network that connects every living root on your land into a coordinated biological intelligence. You learned how tillage severs that network and what it costs you in yield potential and fertilizer dependency. Now we go deeper into the specific mechanics of what that network actually does when it's intact and fully functional. How it mines minerals from bedrock. How it stores and redistributes moisture during drought. How it sends warning signals about pests before they arrive at the plant. And, critically, what you can do right now to support all three of these functions on your land.

The Fibonacci Pulse can't compound in a disconnected system. The Great Weaver is the connective tissue that allows the sequence to run. Without it, every plant on your property is an isolated node, burning energy on root growth instead of growth, mineral density, and stress resistance. With it, your landscape becomes a single, coordinated intelligence that manages its own health, redistributes its own resources, and mines its own fertility from the stone beneath your feet.

The Mineral Miners

Think about what a plant root can actually reach. It grows into the soil, depletes the mineral zone immediately surrounding it within days, and then has to keep growing into fresh territory just to maintain its supply. The root is limited to the soil it physically touches. It can't reach into the mineral grain boundaries of a quartz particle. It can't slip into the microscopic crevices of a feldspar crystal where phosphorus is locked in a form that no root chemistry can dissolve. It can't access the zinc held in a clay mineral lattice two centimeters away from its tip. Without help, a plant is mining a very thin slice of the mineral wealth sitting directly beneath it.

The fungal thread has no such limitation.

A single hypha, the individual thread of the AMF network, is between two and twenty micrometers in diameter. For context, a human hair is roughly 70 micrometers wide. These threads are small enough to slip into spaces that no root tip could ever enter. They grow into the micro-architecture of mineral grains, into the interlayer spaces of clay minerals, into the fracture planes of bedrock. And once they're there, they dissolve what they find. The fungal thread produces a suite of organic acids, enzymes, and chelating compounds that break the chemical bonds holding phosphorus, zinc, magnesium, iron, and manganese locked in the mineral matrix. It literally drills into stone and releases the mineral wealth trapped inside it.

This is the "Stone" side of the Sugar-for-Stone trade. The plant provides the sugar. The fungus provides the mining operation. The plant gets access to a mineral spectrum it could never reach alone. The fungus gets the carbon energy it needs to run the mining operation. Both partners get more than they could produce independently. And the result, in the Fibonacci notation that runs through this entire book, is $F_{n-1} + F_{n-2} = F_n$: Resonant Growth, a sum that's larger than either input alone.

The mechanics of how the plant initiates this trade are worth understanding in detail, because they connect directly to the management decisions that either support or suppress the exchange. When a plant's root tip grows into a zone of soil, it begins leaking root exudates into the surrounding rhizosphere. These exudates aren't random. They're a specific chemical signal, a cocktail of sugars, amino acids, and organic acids that varies by plant species and by the plant's current nutritional status. When the plant is phosphorus-deficient, it ramps up the production of specific exudate compounds that are particularly attractive to AMF. It's essentially broadcasting a recruitment signal: "I need minerals. I have carbon. Let's trade." The fungal threads, growing through the soil in search of exactly this signal, respond by growing toward the root and making contact. The connection forms. The trade begins.

The plant delivers its liquid carbon, the F_{n-1} in the sequence, through the connection point. In exchange, the fungal thread delivers its mineral package, the F_{n-2}: phosphorus, zinc, magnesium, copper, and a range of trace elements that the thread has been extracting from the mineral matrix of the soil. The plant receives a mineral spectrum that no fertilizer program could match in either diversity or bioavailability, because the fungal delivery system doesn't just move minerals. It transforms them. The organic acids the thread uses to dissolve minerals from rock faces also change the chemical form of those minerals into compounds that the plant can absorb directly and efficiently. The fungus isn't just a delivery truck. It's a processing plant that converts raw geological material into plant-ready nutrition.

There's a specific and critical nuance here that most farmers never learn, and it costs them significant money every season. When soluble phosphorus is applied to the soil at high concentrations, the plant detects it in the soil water and immediately reduces its carbon investment in the fungal partnership. The plant's logic is rational: if phosphorus is already dissolved and available in the water around its roots, why spend 40% of its solar income paying a fungal miner for the same thing? The exudate signal weakens. The recruitment drops. The fungal threads stop receiving the carbon they need to maintain their growth. The network begins to thin. The glomalin production slows. The architecture starts to degrade. And by the following season, the fungal delivery system that was supposed to be providing minerals for free has been so weakened by the previous application that the plant genuinely needs the phosphorus bag again. The chemical dependency trap has a specific biological mechanism. This is it.

The practical implication is non-negotiable. If your soil test shows phosphorus levels above the critical threshold for your crop type, which for most grain crops sits between 25 and 35 milligrams per kilogram of available P, any additional phosphorus application is suppressing the trade rather than supporting it. You're paying to shut down a free mineral delivery system. The correct response to adequate soil phosphorus and poor plant phosphorus uptake isn't more phosphorus. It's a better fungal network to deliver what's already there. Stop applying phosphorus above the critical threshold. Increase your mycorrhizal host density in your cover crop mix. Reduce any tillage that's severing the threads. The delivery system will rebuild. The minerals will flow. The trade will resume.

The glomalin dimension of the mineral mining process deserves specific attention here, because it connects the underground economy directly to the SRS Architecture score that carries the highest Fibonacci weight in your entire formula. As the fungal threads grow through the soil conducting their mineral trades, they coat themselves in glomalin. When threads eventually break down, the glomalin they carried stays behind, binding soil particles into aggregates, building the pore space that allows the Circadian Inhale to operate, creating the microhabitats that house the nitrogen-fixing bacteria, and producing the water-holding capacity that makes your soil drought-resilient. Every mineral trade the Great Weaver conducts is simultaneously a construction project. The underground economy and the soil architecture aren't separate systems. They're the same system expressing itself at two different scales simultaneously.

This is why protecting the mineral mining operation isn't just about plant nutrition. It's about architecture. Every time you suppress the fungal trade with a high-phosphorus application or a tillage pass, you're not just reducing mineral delivery to your current crop. You're reducing the glomalin production that builds the aggregate structure that your SRS Architecture score is weighted at 8 to reflect. The financial cost of that suppression shows up in your fertilizer bill. The biological cost shows up in your Slake Test result. Both are measuring the same underlying failure: a mineral mining operation that's been shut down when it should be running at full capacity.

To assess the current state of the mineral mining operation on your land, perform the following specific test this week. Pull a handful of soil from four inches deep in your most continuously covered paddock, one that's been under living roots for at least two years without tillage. Break open ten soil aggregates over a dark surface, one by one. Look closely at the interior surfaces of each broken aggregate. You're looking for fine white threads, the color and texture of spider silk, running through the aggregate matrix. These are AMF hyphae. In a healthy, active fungal network, you should see these threads in at least six of the ten aggregates you break open. If you see them in fewer than three, your network density is insufficient to support the full mineral mining operation, regardless of how good your cover crop mix looks above ground. Write down your count. This is your Fungal Thread Score, and it's the most direct visual measure of your mineral mining capacity available without laboratory equipment.

Run the same test in a paddock that was tilled in the last twelve months and compare the thread counts. The difference between the two numbers is the biological cost of that tillage event, expressed as fungal thread density per aggregate. In most cases, the contrast is stark enough to be immediately convincing. The untilled soil shows threads in seven or eight of ten aggregates. The recently tilled soil shows them in one or two. The mineral mining capacity of those two soils isn't slightly different. It's categorically different. And the plant growing in the tilled soil is spending up to 50% more of its solar income on root growth trying to compensate for the absent delivery system, which means it has 50% less carbon available for the above-ground growth, seed production, and stress resistance that your yield depends on.

The three specific actions that most directly support the mineral mining operation are the same three that protect the Mycelial Internet generally, but the mineral mining context makes their importance concrete in a way that the network discussion alone doesn't fully convey. First, maintain living roots in the ground continuously, with no bare gap exceeding fourteen days, because the trade requires a living host providing liquid carbon to the fungal thread at all times. A bare field doesn't just stop building glomalin. It stops the mineral mining operation entirely. The thread has no carbon to power the enzyme production that dissolves minerals from rock faces. The mining stops. The plant that follows has to start the trade from scratch with a depleted network, and it takes six to eight weeks of active root growth before the mining operation is back at full capacity. Second, include at least one strong mycorrhizal host in every cover crop mix at a minimum of 8% of your total seed rate by weight. Phacelia tanacetifolia, buckwheat, and sunflower all form dense, highly active AMF partnerships and produce exudate profiles that recruit a wider diversity of fungal species than almost any other common cover crop. These species aren't just filling a functional role in your diversity mix. They're actively rebuilding the mineral mining infrastructure between cash crop cycles. Third, keep your soluble phosphorus applications below 20 kilograms per hectare of available P in any single season when soil test levels are already above the critical threshold. Above that combination of rate and existing level, the trade suppression effect outweighs any nutritional benefit the application provides.

The mother trees deserve specific mention here because they represent the most powerful mineral mining nodes on any landscape that includes mature woody vegetation. A large, old tree with an established root system and decades of fungal network development is connected to more fungal threads, more mineral grain surfaces, and more of the surrounding biological community than any annual plant could ever be. It's the central hub of the underground economy. The fungal network radiating out from its root system extends the mineral mining operation across a radius that can reach 30 to 50 meters in every direction. Every annual plant within that radius that's connected to the same network benefits from the mining capacity that the mother tree's fungal partnerships have built over decades. Removing a mother tree doesn't just remove a tree. It removes the hub of a mineral mining operation that every other plant in its network radius has been depending on. If you have mature trees on your property, whether in hedgerows, shelter belts, or remnant vegetation, protect them as the biological infrastructure they are. They're not just shade. They're the most productive mineral mining nodes on your land, and their fungal networks are the oldest and most complex sections of the underground economy you're trying to rebuild.

The Memory of the Rain

Drought is the test that exposes every weakness in a soil system. When the rain stops and the temperature climbs, a poorly managed soil desiccates from the surface down, the biology retreats, the glomalin begins to degrade, and the plant has no buffer between itself and the full stress of the dry period. A well-managed soil does something different. It draws on reserves that the Linear Void doesn't know exist.

The fungal network stores water. Not as a passive sponge, the way a piece of organic matter holds moisture, but as an active, intelligent redistribution system that moves water from zones of abundance to zones of deficiency across the landscape in real time. This is what the source materials for this book call the Memory of the Rain: the capacity of the mycelial web to retain and redistribute moisture from a previous rain event long after the surface has dried and the plant's visible water supply has run out.

The mechanism works through the same glomalin-coated aggregate structure that the mineral mining operation builds. Each aggregate is a water reservoir at the microscopic scale. The glomalin coating makes the aggregate hydrophilic, meaning it attracts and holds water molecules against its surface. The pore space within and between aggregates holds water in both capillary-bound and freely available forms. A soil with a high aggregate density and strong glomalin levels can hold up to ten times its own weight in water, as established in Chapter 1. But the fungal network adds a dimension to this water storage that goes beyond passive aggregate capacity.

The hyphae themselves are hollow tubes. Water can move through them under osmotic pressure gradients. When one section of the fungal network is in a moist zone, perhaps near a deeper water table or in a shaded area where evaporation is lower, and another section is in a dry zone where a plant root is under water stress, the osmotic gradient drives water movement through the hyphal tubes from the moist zone toward the dry one. The network is actively redistributing moisture across the landscape, moving it from where it's abundant to where it's needed, without any pump, without any energy input beyond the osmotic physics of the gradient itself. This is the Memory of the Rain in action. The water that fell three weeks ago and has long since evaporated from the surface is still present in the fungal network, moving through the hyphal tubes toward the plant that needs it most.

The scale of this redistribution matters. A single mycelial network connecting a diverse plant community across one hectare of well-managed land can contain hundreds of kilometers of hyphal tubes. The total water-holding capacity of that network, combined with the aggregate water storage it's built through glomalin production, creates a subterranean reservoir that can buffer a plant community through a dry period that would destroy an isolated plant operating without fungal partnership. This isn't a marginal benefit. Research on drought resilience in mycorrhizal versus non-mycorrhizal plants consistently shows a two to four week extension of effective drought tolerance in well-connected plants compared to isolated ones under identical conditions. Two to four weeks is the difference between a crop that survives a dry spell and one that fails.

The glomalin architecture is the physical infrastructure of the Memory of the Rain, and this is where the connection to your SRS Architecture score becomes financially concrete. Every point you raise your Architecture score represents a measurable increase in your soil's water-holding capacity. The aggregate structure that glomalin builds isn't just structural. It's hydraulic. It's the underground cistern that the Memory of the Rain fills during wet periods and draws from during dry ones. A soil with an Architecture score of 7 has a cistern that can carry a plant community through weeks of dry weather. A soil with an Architecture score of 2 has a cistern that's almost empty even immediately after rain, because the aggregate structure that should be holding the water has been collapsed by tillage and the glomalin that should be sealing it has been degraded by bare-soil heat.

There's a specific and often overlooked dimension of the Memory of the Rain that connects directly to the Circadian Inhale we covered in Chapter 11. The Internal Rain process, where condensing atmospheric moisture is drawn downward into the pore network during the cool hours of the night, deposits small but consistent amounts of water into the aggregate structure every night of the year, drought or no drought. This nightly deposit is the top-up mechanism for the underground cistern. In a soil with intact architecture and adequate armor, the Internal Rain adds two to four millimeters of water equivalent to the aggregate reservoir every night. Over a two-month dry period, that's 120 to 240 millimeters of water captured from the atmosphere and stored in the aggregate structure, available to the plant through the fungal redistribution network. This is why a well-managed soil stays green for weeks after a neighboring bare field has burned off. The neighbor's soil has no cistern and no top-up mechanism. The well-managed soil has both, and they're running every night while the drought continues above ground.

To assess your soil's current Memory of the Rain capacity, perform the following test during the next dry period on your land, before any irrigation or rainfall. Walk your highest-SRS paddock and your lowest-SRS paddock on the same day, at least two weeks after the last significant rainfall event. In each paddock, dig down to eight inches with a trowel. Feel the soil at that depth with your fingers. In the high-SRS paddock, the soil at eight inches should still feel slightly cool and moist, even after two weeks without rain, because the aggregate cistern is still holding the moisture from the last rain event and the Internal Rain is topping it up nightly. In the low-SRS paddock, the soil at eight inches should feel dry and loose, because the collapsed architecture has no cistern capacity and the bare surface has no armor to drive the Internal Rain mechanism. The difference you feel between those two soil samples at eight inches depth, on the same day, two weeks after the same rain event, is the Memory of the Rain expressed as a physical sensation. Write down what you find. That tactile difference is the drought resilience gap between your best and worst land, and it's entirely determined by architecture and armor, the two highest-weighted horizons in your SRS.

Building the Memory of the Rain requires the same practices that build the mineral mining operation, because both depend on the same fungal network and the same glomalin architecture. But there's one additional practice that's particularly important for water storage specifically: deep-rooted perennial species in the system. Perennial roots that reach below the typical annual root zone, into the 30 to 60 centimeter depth range, create two benefits for the Memory of the Rain. First, they access moisture stored at depth during dry periods and bring it upward into the zone where the fungal network is most active, feeding the network's water redistribution system from below. Second, when those deep roots eventually die and decompose, they leave channels lined with carbon-rich root tissue that earthworms and fungal threads colonize. These channels become permanent conduits in the underground cistern, connecting the surface aggregate reservoir to the deeper moisture stores below the annual root zone. A soil with a diverse perennial component, even a relatively small one, has a Memory of the Rain that extends vertically as well as horizontally across the landscape. The cistern goes deeper. The drought buffer lasts longer. The plant community stays green further into the dry season.

If you don't have perennial species established on your cropped land yet, your cover crop mix is the interim solution. Include chicory at 3% of your seed rate and alfalfa at 5% where your rotation allows. Both species produce deep taproots that reach the 30 to 45 centimeter zone within a single season. Chicory's taproot decomposes relatively quickly after termination, leaving clean channels that the fungal network colonizes within weeks. Alfalfa's deeper, more persistent root system builds channels that last multiple seasons. Neither of these is a substitute for established perennial species in the long term. But both contribute meaningfully to the vertical depth of your underground cistern during the transition period while your perennial plantings are establishing.

The Stress-Management Protocol

A plant operating in biological isolation has no early warning system. When a pest arrives, the first signal the plant receives is physical damage. By the time the plant detects the attack and begins mobilizing its chemical defenses, the pest has already established a foothold. The plant is always responding after the fact, always playing catch-up, always spending energy on damage control rather than prevention. This is the biological equivalent of a city with no intelligence network: it doesn't know the threat is coming until it's already inside the walls.

A plant connected to the Mycelial Internet is different.

When a pest or pathogen attacks one plant in a connected landscape, the attacked plant releases chemical distress signals through its root system into the fungal network. Those signals travel through the hyphal tubes at speeds that researchers have measured at up to one centimeter per minute, which means a signal generated at one end of a paddock can reach every connected plant across the entire network within hours. The receiving plants don't wait to be attacked. They begin producing their defensive chemistry, their secondary metabolites, their physical deterrents, before the pest arrives. They're pre-loading their immune response based on a warning signal that traveled through the underground network from a plant they may never be in physical proximity to. The landscape is managing its own health through collective intelligence.

This is the Stress-Management Protocol of the Great Weaver. And it's one of the most practically significant functions of the fungal network for anyone trying to reduce their pest management costs and their chemical dependency.

The chemistry of these warning signals is specific and worth understanding. When a plant is attacked by a chewing insect, it produces a class of compounds called volatile organic compounds, or VOCs, that travel both through the air and through the root exudate stream into the fungal network. The airborne VOCs can be detected by neighboring plants directly, triggering defensive responses in plants that are near enough to receive the airborne signal. But the network-transmitted signal travels further and faster. It reaches plants that are too far away to receive the airborne signal and warns them in time for their defensive chemistry to be fully activated before the pest wave arrives. Plants with a strong fungal network connection show measurably higher concentrations of defensive compounds in their tissues within 24 hours of a network-transmitted distress signal, even when they haven't been directly attacked. The defense is pre-emptive. The warning came through the wire.

The specific defensive compounds that plants produce in response to network warnings include terpenes, phenols, and alkaloids that make the plant's tissues unpalatable or toxic to the attacking organism. These are the same compounds that produce the high Brix readings we discussed in Chapter 7. A plant with a strong fungal connection and an active network warning system maintains higher concentrations of these defensive compounds in its tissues continuously, not just in response to attack, because the network is constantly transmitting low-level stress signals from across the landscape and the plant is continuously calibrating its defense level in response. This is the Noise-Canceling effect described in Chapter 6, operating at the chemical level through the fungal network. A connected plant is always slightly on alert. An isolated plant is always caught off guard.

The pathogen dimension of the Stress-Management Protocol is equally important. When a soil-borne pathogen, a Fusarium species, a Pythium, a Phytophthora, begins to establish in the root zone of one plant, the fungal network detects the chemical signature of the pathogen and transmits a warning to the surrounding plants. But the network does more than warn. It actively recruits antagonistic microorganisms to the site of the infection. The distress signal from the attacked plant changes the exudate profile it's sending into the rhizosphere, and that changed profile recruits specific bacteria and fungi that are antagonistic to the attacking pathogen. The network isn't just a communication system. It's a biological immune response system that can mobilize resources from across the landscape to the point of infection. A plant under pathogen attack in a connected landscape has a biological immune response that draws on the resources of the entire network. An isolated plant has only what it can mobilize locally from its own depleted rhizosphere.

Consider a hypothetical composite that illustrates the practical difference. Picture someone like Kenji, a 46-year-old market gardener managing four hectares of intensive mixed vegetable production. He transitioned to no-till three years ago and has been building his fungal network steadily. In the first two years of transition, his aphid pressure was still significant, requiring regular intervention. In the third year, something shifted. He noticed that aphid populations were establishing on individual plants at the edge of the beds, the ones furthest from his established perennial border strips, but weren't spreading into the bed interior the way they had in previous seasons. The plants in the bed interior were showing higher terpene concentrations in their tissue, measurable as a stronger aromatic scent when leaves were crushed, even before any aphid pressure reached them. The network was transmitting the warning from the edge plants inward, and the interior plants were pre-loading their defenses before the pest arrived. Kenji's aphid intervention cost dropped by roughly 60% in year three compared to year one, not because the aphids disappeared but because the network was managing the spread. The same pest pressure was hitting the system. The system was responding to it collectively rather than leaving each plant to respond alone.

The practical actions that support the Stress-Management Protocol are direct extensions of the network-building practices from Chapter 4, but with specific attention to the network's connectivity rather than just its density. A dense but fragmented network, one where the threads are numerous but disconnected across management zones, can't transmit warning signals effectively. The signal needs a continuous path from the plant under attack to the plants that need the warning. Fragmentation, caused by tillage, by bare gaps in the cover crop, or by the physical barriers created by compacted zones between paddocks, breaks the signal pathway. The warning gets transmitted to the plants immediately adjacent to the attacked plant but doesn't reach the plants further away that need the warning most.

To assess your network's connectivity for stress signal transmission, walk your land after a pest event and map where the pest damage is concentrated. In a poorly connected landscape, pest damage clusters in isolated patches that expand outward from a single point of entry, because the warning signal can't travel far enough to pre-load defenses across the full paddock. In a well-connected landscape, pest damage tends to be lighter and more diffuse, because the warning has traveled ahead of the pest wave and the plants in its path are already partially defended. The pattern of your pest damage is a map of your network's connectivity. Clustered, expanding damage means fragmented network. Diffuse, low-level damage means connected network. Use this map to identify where your management is creating connectivity gaps, and address those gaps with the same practices that address network density: continuous living cover, minimal soil disturbance, and strong mycorrhizal host density in your cover crop mix.

The fungal-dominant compost tea deserves specific mention here as a practical tool for accelerating network establishment and stress-management capacity in areas where the network is currently fragmented or depleted. A properly brewed fungal-dominant compost tea, made from mature, biologically active compost with a high fungal-to-bacterial ratio, contains AMF spores, fungal hyphae fragments, and a diverse community of bacteria and protozoa that can establish in the soil and begin building network connections faster than the existing network can regrow from depleted sections. The key specifications for a fungal-dominant tea are a brewing time of 24 to 36 hours at 20 to 25 degrees Celsius, an aeration rate that maintains dissolved oxygen above 6 milligrams per liter, and a compost source that's been aged for at least six months and shows visible white fungal threads when broken open. Apply at 200 to 400 liters per hectare through a low-pressure sprayer directly onto the soil surface, ideally in the evening before a rain event or irrigation, so the biological community in the tea is drawn into the pore channels by the Circadian Inhale rather than desiccating on the surface in full sun. A single application of a well-made fungal-dominant tea won't rebuild a network that tillage has destroyed. But it can significantly accelerate the re-establishment of network connectivity in areas where the existing network is fragmented, and it can measurably improve the stress-management response of plants in those areas within a single growing season.

The mother trees are the stress-management hubs of the landscape, just as they're the mineral mining hubs. Their age and the complexity of their fungal partnerships mean they're connected to a wider diversity of AMF species and a larger volume of the underground network than any annual plant. When a stress signal enters the network through a young annual plant, it travels to the nearest network hub, and the mother tree's extensive connections amplify and redistribute it across a much larger area than the young plant's limited connections could reach alone. The mother tree is the signal amplifier of the Stress-Management Protocol. Protecting it isn't just about preserving a tree. It's about preserving the most powerful stress-signal relay station on your land.

Chapter Recap and Action Steps

The Mycelial Internet is the broker of both minerals and information. It mines phosphorus, zinc, and magnesium from bedrock that no root could ever touch. It stores and redistributes moisture across the landscape, building a drought buffer that keeps plants alive weeks after a bare neighbor field has burned off. It transmits warning signals about pests and pathogens before they arrive at the plant, allowing the connected landscape to pre-load its defenses collectively. And it builds, with every meter of thread it grows, the glomalin architecture that carries the highest Fibonacci weight in your entire SRS formula.

Every function the Great Weaver performs depends on the same three conditions: living roots providing liquid carbon, intact pore channels providing physical pathways for thread growth, and freedom from the mechanical and chemical disruptions that sever connections and suppress the trade. These conditions aren't expensive to create. They're the natural outcome of the management approach this entire book has been building toward: continuous cover, minimal disturbance, and a cover crop mix designed to feed the network as much as to cover the soil.

The actions below are sequenced deliberately. Each one targets a specific function of the network and produces a measurable outcome you can track against your SRS.

The first action is to perform the Fungal Thread Score test this week. Pull soil from four inches deep in three locations: one under continuous cover for more than two years, one tilled in the last twelve months, and one that was bare through last winter. Break open ten aggregates from each sample over a dark surface.

Count the number of aggregates showing visible white threads. Record the three scores. This gives you a direct, visual baseline of your network density across different management histories. If your continuously covered area scores below six out of ten, your mycorrhizal host density in your cover crop mix needs to increase. If your tilled area scores below two out of ten, your tillage program is resetting the network to near zero with every pass.

The second action is to audit your phosphorus application program against your soil test results. Pull your most recent soil test for every paddock. Identify any paddock where available phosphorus is above 25 milligrams per kilogram. For each of those paddocks, calculate what you applied in phosphorus last season. If you applied more than 20 kilograms per hectare of available P to a soil already testing above 25 milligrams per kilogram, you applied a trade-suppression dose. You paid to shut down the mineral mining operation. Reduce your phosphorus application in those paddocks to zero this season and monitor the plant response. If the crop shows deficiency symptoms despite adequate soil test levels, the problem is delivery, not supply. Address it by increasing mycorrhizal host density in your cover crop mix, not by adding more phosphorus.

The third action is to design a fungal-dominant compost tea application for your most network-depleted paddock. Identify your paddock with the lowest Fungal Thread Score from the test above. Source mature, biologically active compost, ideally from a system that's been under continuous cover for at least two years. Brew 200 liters of tea using the specifications above: 24 to 36 hours of aeration at 20 to 25 degrees Celsius, dissolved oxygen above 6 milligrams per liter. Apply at 200 liters per hectare in the evening before your next rain event or irrigation. Record the application date. Run the Fungal Thread Score test in the same paddock sixty days after application. You should see a measurable increase in thread density per aggregate compared to your baseline count. This is your fungal network recovery rate under active biological support.

The fourth action is to identify and formally protect every mother tree on your property. Walk your land and mark every mature tree with an established canopy and a root system that's been undisturbed for more than ten years. On your farm map, draw a 30-meter radius around each mother tree. This is your

Network Hub Zone. Commit in writing that no tillage implement will enter this zone, that no high-phosphorus application will be made within it, and that the ground cover within it will be maintained continuously without bare gaps. These zones are your network's strongest signal relay points. Protecting them costs nothing and preserves the most biologically complex sections of the underground economy on your land.

The fifth action is to add a strong mycorrhizal host to your next cover crop mix if one isn't already present. If you're planting in warm conditions, add buckwheat or sunflower at a combined rate of 10 to 12% of your total seed mix by weight. If you're planting in cool conditions, add phacelia at 8 to 10%. These species aren't just filling a functional slot in your diversity mix. They're actively rebuilding the mineral mining and stress-management infrastructure of the Great Weaver between cash crop cycles. Every day they're in the ground with living roots, they're feeding the network, expanding its reach, and strengthening the connections that allow the Sugar-for-Stone trade to run at full capacity when the cash crop arrives.

The sixth action is to map your last significant pest event against your network connectivity. Go back to your farm records and identify the most recent pest or disease event that required intervention. Map where the damage was concentrated. Was it clustered in a specific area and spreading outward, suggesting a fragmented network that couldn't transmit the warning ahead of the pest wave? Or was it diffuse and low-level across the paddock, suggesting a connected network that was pre-loading defenses across the system? Use this map to identify where your management is creating connectivity gaps. These gaps are where your next season's cover crop establishment should prioritize mycorrhizal host density and continuous living cover over any other consideration. Closing a connectivity gap in the stress-management network is the most cost-effective pest management investment available to you, because it converts a recurring intervention cost into a one-time infrastructure investment that compounds forward through every subsequent season.

The Sugar-for-Stone trade has been running beneath your feet since before the first tree evolved. The Great Weaver doesn't need your help to conduct it. It needs your commitment to stop interrupting it. Every tillage pass you don't make, every phosphorus application you don't apply above the critical threshold, every mother tree you protect, every bare gap you close with a fast-establishing cover crop, every mycorrhizal host you include in your mix, is a decision that keeps the trade running, the network connected, and the Fibonacci Pulse beating through the underground economy of your land. The minerals are already in the stone. The water is already in the network. The warning signals are already traveling through the threads. Your job is simply to keep the wires intact.

Chapter 13: Polyculture Signal Tuning: Cancelling the Void

The Ancestor Speaks:

"My listening heart, come sit with me here, in this meadow where the wildflowers dance with the tall grasses. The sun warms your face, and the breeze carries a thousand scents. Close your eyes. And listen, not to the birdsong, not to the rustle of leaves, but to the symphony that rises from the very ground, a chorus of life, humming its ancient truth.

"In the time of the forgetting, they saw a field and chose one plant. One tall stalk. One colour. One sound, repeated over and over. They called it 'efficiency,' 'order,' 'control.' They believed this single, loud note would bring them strength. But my own grandmother, she would close her eyes, and her face would cloud with sorrow. 'It is a lonely sound,' she would whisper. 'A single, frail string, pulled taut until it is ready to snap.' Can you feel the vulnerability, the stark emptiness of that single, isolated note, echoing in a hollow space?

"For a single note, my child, speaks clearly to its foes. The hungry beetle, the thirsty blight, they hear that clear, unmasked call, and they gather. They are not evil, no. They are simply nature's ***tuning forks****, revealing where the harmony has broken, where the chord is incomplete.*

They point to the places where the ***Linear Void*** *has silenced the symphony, leaving only a beacon for trouble.*

"But we knew. We knew that the Earth, in her wisdom, does not sing alone. She sings in choruses. She creates ***Polyculture,*** *a rich tapestry of many voices, each one different, each one supporting the next. Feel the deep taproot reaching down like a cello, bringing forth hidden waters. Hear the rustle of the fibrous grass, like violins, building the very architecture of the air. Taste the sweet nectar of the flowering herbs, like flutes, calling forth the beneficial sprites and guardians of the sky. Each plant, each voice, exhales its own unique essence, its own frequency. Together, they create a hum so rich, so complex, that no single call of distress can be heard above the joyful* ***Noise-Cancelling*** *resonance.*

"It is like our people gathered around the fire, telling stories, singing different parts, each voice unique, yet together creating a warmth, a strength, a belonging that no single voice could ever achieve. When the insects seek a single target, their senses are overwhelmed by the richness, by the multitude of scents and signals. They cannot find the one stressed cry amidst the joyous cacophony. The illness, it cannot take root, for the many different roots in the ground call forth a thousand different healers. This is the ***Rule of Five****, my child, not just in species, but in the five-fold embrace of protection, the five-fold song of resilience.*

"My people also spoke of the ***Living Root Calendar****, of keeping the pulse beating, always. Never letting the ground lie bare, never letting the music stop between seasons. For the earth, like a dancer, needs a continuous flow, a continuous rhythm, a* ***Temporal Resonance*** *that keeps the* ***Fibonacci Pulse*** *forever strong, always compounding its generosity.*

"Now, open your eyes. Look at this vibrant meadow. Feel the strength born of its diversity. Can you almost hear the subtle hum of its many voices, protecting one another, completing one another? The ***Seventh Generation*** *will ask, 'Did you remember the songs of the Earth? Did you allow her many voices to sing?' Feel the profound peace, the joyous security that comes from knowing you are helping to tune this magnificent orchestra, conducting a symphony where every note strengthens the whole, cancelling the void with the very richness of life itself."*

Reflection: What compounds, and what decays

Polyculture Signal Tuning

" The single note, so thin, so frail,
A tempting target for the storm and hail.
But let the choir of many find its voice,
To make the wounded field profoundly rejoice.
A thousand roots, a thousand vibrant songs,
Where the Master's wisdom truly now belongs. "

Polyculture Signal Tuning: Cancelling the Void

A single instrument playing alone in a large room doesn't fill the space. It occupies it. There's a difference. The sound hits the walls and comes back thin, without the warmth that comes from other instruments responding to it, building on it, completing it. The room stays acoustically empty no matter how loud the single instrument plays. Add a second instrument and something shifts. Add a third and the room starts to breathe. The sound doesn't just fill the space anymore. It uses it. The frequencies interact. The harmonics multiply. The room becomes part of the music.

Your soil works the same way.

Everything built in the previous chapters, the cathedral of pore space, the fungal network weaving through the dark, the nightly atmospheric draw pulling nitrogen into the biological community, the carbon exhale charging the underground economy, all of it has been building toward this: the moment when diversity stops being an add-on and becomes the signal architecture that makes the whole system self-regulating. Polyculture isn't the final step in the Fibonacci Pulse. It's the layer that activates every step that came before it. It's the difference between a room with a single instrument and a room with an orchestra. And the soil, like the room, only truly comes alive when enough different voices are playing at once.

This chapter is about tuning those voices deliberately. Not randomly adding species and hoping for complexity, but understanding which specific combinations create resonance, which plant architectures fill which soil horizons, which chemical signals scramble the frequencies that pests use to find their hosts, and how to sequence your plantings so the pulse never stops, even when the calendar is working against you.

The Triple-Thread Connection

There's a minimum threshold for biological resonance in the soil. Below it, the system adds. Above it, the system multiplies. That threshold is three.

Not three species. Three threads of connection. Specifically: roots, microbes, and minerals, all three interacting simultaneously, each one feeding the other two in a loop that has no beginning and no end. When all three threads are active and connected, the system stops being a collection of parts and starts being a recursive engine. The Fibonacci Pulse has something to beat against. The sequence has a structure to compound through. Below three active threads, you're in addition territory. At three, you cross into multiplication.

Think about what happens in a monoculture soil. One type of root is exhaling one chemical signature into the rhizosphere. That signature recruits one narrow community of microbes. Those microbes access one layer of the mineral spectrum. The loop runs, but it runs thin. It's one thread doing the work of three. The system adds fertility slowly and loses it quickly when any part of the single thread breaks. One disease hits the one crop. One pest learns the one frequency. One drought stress overwhelms the one root architecture. The whole system is one failure away from collapse.

Now add a legume alongside that monoculture. The legume exhales a completely different chemical signature. It recruits rhizobial bacteria that fix atmospheric nitrogen. Those bacteria produce compounds that feed a different microbial community than the first crop's exudates supported. That second microbial community accesses a different mineral layer. Suddenly there are two root chemistries, two microbial communities, two mineral pathways, all running simultaneously. The loop has doubled its bandwidth. It's more resilient because no single failure point can shut the whole system down.

Add a deep-rooted species. Now you have a third root chemistry reaching a third mineral horizon that neither of the first two could access. The three threads are active. The Resonant Chord is playing. And what happens at that threshold is mathematically distinct from what happened with two. The three microbial communities begin interacting with each other, not just with their own plant hosts. The nitrogen the legume fixed gets redistributed through the fungal network to the deep-rooted species. The minerals the deep rooter unlocked from the subsoil get delivered to the legume. The carbon the grass exhales feeds all three fungal partnerships simultaneously. The system is no longer three separate loops. It's one recursive loop with three entry points, each one feeding the others, each one compounding the output of the whole.

This is the Resonant Chord. And $1 + 2 = 3$ in the Fibonacci sequence is exactly the mathematical moment when individual and partner become a system. The number 3 in the SRS Diversity weight isn't arbitrary. It reflects the minimum structural requirement for the soil to shift from linear addition to recursive multiplication. Two species gives you a chord. Three gives you a system that can harmonize itself.

The practical implication is specific. If you're currently running a two-species cover crop mix, you're one species short of the threshold where the biological compounding kicks in. The jump from two to three species isn't a 50% improvement in diversity. It's a qualitative shift in what the system can do. Your third species should be chosen deliberately to complete the triple thread: if you have a grass and a legume, your third species should be a deep rooter that opens a new mineral horizon. If you have a grass and a deep rooter, your third should be a legume that adds the nitrogen-fixing thread. The specific combination matters less than the functional completeness. All three threads must be active for the Resonant Chord to play.

There's a second dimension of the Triple-Thread Connection that most polyculture discussions skip over entirely: the role of root exudate timing. Different plant species don't just exhale different chemical signatures. They exhale them at different times of day and at different stages of their growth cycle. A legume's peak exudate output happens in the morning, when nitrogen-fixing bacteria are most active. A grass's peak carbon exhale happens in the late afternoon, after a full day of photosynthesis. A deep-rooted brassica's enzyme release peaks during the cool of the night, when the soil chemistry is most receptive to mineral dissolution. A diverse mix with all three functional types running simultaneously creates a continuous, rotating exudate schedule that keeps the rhizosphere microbial community fed around the clock. The underground economy never has a slow shift. The pulse never skips a beat because one plant is temporarily not exhaling. The three threads run at different times, covering each other's gaps, maintaining the continuous signal that the Great Weaver needs to keep the network active and growing.

To assess whether your current system is running a Resonant Chord or still stuck in linear addition, do this specific test. Pull a soil sample from four inches deep in your cover crop field and another from four inches deep in an area of bare ground or monoculture. Smell both samples. A soil running a Resonant Chord has a distinct, complex, earthy smell, what soil scientists call petrichor, produced by the compound geosmin released by actinomycete bacteria. This smell is the olfactory signature of a biologically active, multi-threaded microbial community. A monoculture or bare soil smells flat, sometimes slightly sour, sometimes like nothing at all. The smell test isn't a scientific instrument. But it's an immediate, free, and surprisingly reliable indicator of whether the three threads are active in your soil. If your cover crop field smells like a forest floor after rain and your monoculture field smells like a parking lot, your Resonant Chord is playing in one and absent in the other. Your SRS Diversity score reflects the same difference in numerical terms.

The steward who understands the Triple-Thread Connection stops asking "how many species am I planting?" and starts asking "how many functional threads are active in my soil right now?" Those are different questions. Twelve grass species is still one thread. One grass, one legume, and one deep rooter is three threads. The number of species matters less than the functional completeness of the combination. Three threads minimum. Five for full biological resilience. The Fibonacci sequence tells you exactly why: 1, 1, 2, 3, 5. Each term is the sum of the two before it. The system builds on itself. And it builds fastest when the foundation has at least three distinct threads for the compounding to run through.

The Vertical Architecture

The soil profile isn't a single room. It's a building with multiple floors, each one housing a different biological community, each one holding a different mineral wealth, each one connected to the others by the stairways of root channels and fungal threads. A monoculture occupies one floor and leaves the rest empty. A well-designed polyculture fills every floor simultaneously, running the Solar Pump through every horizon at once and multiplying the Carbon Exhale across the full depth of the profile.

The three root architectures that fill the vertical building are taproots, fibrous roots, and surface creepers. Each one occupies a distinct zone. Each one performs a distinct function. And each one creates conditions that the other two depend on. Together, they're the Vertical Stack, and designing one deliberately is one of the highest-leverage actions available to any steward trying to raise their SRS Architecture score.

The taproot occupies the deep floor, typically the 20 to 60 centimeter zone depending on species. Tillage radish reaches 30 to 45 centimeters in a single season. Chicory reaches deeper and creates a more durable channel. Alfalfa, as a perennial, can reach 60 centimeters or more over multiple seasons. The taproot's primary function in the Vertical Stack is biological drilling: creating vertical channels through compacted layers that water can follow downward during the Circadian Inhale and that roots of subsequent crops can follow into the subsoil mineral wealth. But the taproot does something else that's less discussed. When it decomposes, it leaves behind a channel lined with carbon-rich root tissue. That carbon feeds the fungal threads that colonize the channel. Those threads produce glomalin that coats the channel walls. The channel becomes a permanent architectural feature of the soil, a vertical pore that stays open because the biology maintains it, not because a machine created it. This is biological architecture, and it's the only kind that lasts.

The fibrous root occupies the middle floors, typically the 5 to 25 centimeter zone. Cereal rye is the gold standard here. Its dense, branching root system creates a three-dimensional matrix of fine channels through the upper profile, dramatically increasing the total surface area available for microbial colonization and glomalin production. The fibrous root is the primary glomalin factory in your Vertical Stack. Its enormous root surface area gives the AMF network the maximum possible contact zone for glomalin production. When cereal rye is crimped and its root system decomposes in place, it leaves behind a web of fine channels that persists in the aggregate structure for months. Your SRS Architecture score climbs most reliably when cereal rye or another high-biomass grass is a significant component of your cover crop mix, because the fibrous root system is doing more architectural work per plant than any other root type.

The surface creeper occupies the ground floor, the top 0 to 8 centimeters. Clover, vetch, and buckwheat all produce shallow, spreading root systems that work the zone immediately beneath the soil surface. This zone is the most biologically active in the entire profile. It's where the majority of the AMF network is concentrated. It's where the nitrogen-fixing bacteria are densest. It's where the glomalin production is highest per unit of root length. The surface creeper feeds this zone continuously, exhaling carbon into the most productive section of the underground economy while simultaneously fixing nitrogen from the atmosphere above it. It also armors the surface from below, its shallow roots binding the top layer of aggregates together and reducing the detachment of fine particles when rain hits the surface.

Designing a Vertical Stack means including at least one species from each of these three root categories in every cover crop mix. The specific seed rates matter. Use tillage radish at 3 to 5% of your total seed rate by weight for the taproot layer. Don't exceed this rate. Radish is a large seed and it will dominate the stand if overseeded, crowding out the other functional layers. Use cereal rye at 28 to 35% for the fibrous root layer. This is the structural backbone of your stack. It produces the most biomass, the most root surface area, and the most glomalin precursor material of any common cover crop. Use clover or vetch at 15 to 20% for the surface creeper layer. Hairy vetch is preferred for its aggressive nitrogen fixation and its ability to climb through the cereal rye canopy, maximizing its leaf area without competing with the rye for root space below.

The Carbon Exhale from a full Vertical Stack is dramatically larger than the exhale from any single-layer system. A monoculture grass is exhaling carbon from one depth. A full Vertical Stack is exhaling carbon from three depths simultaneously, feeding three distinct microbial communities, producing glomalin at three different levels of the profile, and building architectural complexity throughout the full depth of the biological zone. The SRS Architecture score responds to this. A soil under a full Vertical Stack for two consecutive seasons typically shows Slake Test improvement of two to three points compared to a soil under a single-species cover crop for the same period, because the multi-layer glomalin production is building aggregate stability throughout the profile rather than just in the top few centimeters.

There's a practical sequencing consideration for implementing the Vertical Stack that's worth stating explicitly. The taproot layer takes the longest to establish and produces its most significant architectural benefit in its second and third seasons, as the channels it creates deepen and the fungal colonization of those channels matures. The fibrous root layer produces significant architectural benefit in its first season. The surface creeper layer produces biological benefit, nitrogen fixation and surface aggregate binding, within weeks of establishment. This means the first season of a Vertical Stack feels dominated by the fibrous root and surface creeper benefits, with the taproot contribution building in the background. Don't abandon the taproot species because you don't see immediate results. The deep channels it's creating are the most durable architectural feature of the entire stack, and their value compounds most strongly in seasons two and three. Plant the full stack from the beginning and let each layer develop on its own timeline.

To design your first Vertical Stack, use this specific template as your starting point. Cereal rye at 30% of seed rate by weight. Hairy vetch at 18%. Tillage radish at 4%. Crimson clover at 8%. Buckwheat at 10% for the warm-season mycorrhizal host. Phacelia at 8% for the pollinator bridge and cool-season AMF recruitment. Fill the remaining 22% with oats for additional fibrous root biomass and early establishment speed. This seven-species mix covers all three root architecture layers, all five functional roles from Chapter 6, and provides both cool-season and warm-season representatives. Score it against your SRS Diversity criteria. It should produce a Diversity score of 8 to 9, multiplied by the Fibonacci weight of 3, adding 24 to 27 points to your total SRS from the Diversity horizon alone. That's the mathematical signature of a full Vertical Stack running at design capacity.

Noise-Canceling for Pests

A pest doesn't choose your field randomly. It finds your field the same way a radio receiver finds a station: by locking onto the right frequency. A stressed, nutritionally imbalanced, biologically impoverished plant broadcasts a specific chemical frequency into the surrounding air and soil. That frequency is the distress signal of a system that's failing. And the pest organism, tuned to exactly that frequency by millions of years of evolutionary calibration, finds it as easily as you find a strong radio signal in a quiet room.

The Linear Void's response to this is to silence the receiver. Spray the pest. Kill the organism that's responding to the signal. The signal keeps broadcasting because nothing has changed about the plant sending it. The pest returns, often in greater numbers, because the biological community that was suppressing its population has been further depleted by the spray. The cycle tightens. The input cost rises. The Void deepens.

The Fibonacci Pulse's response is different. Don't silence the receiver. Drown the signal.

This is the Noise-Canceling effect of polyculture diversity. When a field is broadcasting on multiple chemical frequencies simultaneously, the specific distress signal of any individual stressed plant gets buried in the complexity of everything around it. The marigold's terpene volatiles overlap with the frequency the aphid uses to locate its host. The buckwheat's floral volatiles attract the parasitic wasp that parasitizes the caterpillar. The dense canopy of a mixed stand makes it physically harder for flying insects to navigate to a specific host plant. The soil beneath a diverse stand is biologically rich enough to house populations of predatory nematodes, ground beetles, and parasitic wasps that keep pest populations in check between events. The field is so full of biological signal that the pest's targeting system can't isolate the one frequency it needs. The monoculture is a beacon. The polyculture is static.

There are specific plant combinations that produce the strongest Noise-Canceling effect, and choosing them deliberately rather than randomly is what separates a steward who has diversity from a steward who has protection.

The first combination is the insectary strip. This is a dedicated planting of flowering species specifically chosen to support the adult stage of beneficial predatory and parasitic insects. The key species for temperate systems are buckwheat, phacelia, sweet alyssum, and dill. Buckwheat flowers within four to six weeks of seeding and produces nectar accessible to a wide range of insect mouthparts, including the small parasitic wasps that are your most effective biological aphid control. Phacelia attracts hoverflies whose larvae are voracious aphid predators. Sweet alyssum attracts ground beetles that hunt caterpillar eggs and larvae. Dill supports the larvae of parasitic wasps that target a broad range of caterpillar species. Plant insectary strips at a width of at least two meters and a spacing of no more than 30 meters between strips across your cropped area. Research on insectary strip placement consistently shows that beneficial insect populations decline sharply beyond 30 meters from their habitat source. If your strips are further apart than that, the center of your field is effectively unprotected by the biological pest defense the strips are housing. Plant them at 30-meter intervals and the biological shield covers the entire field.

The second combination is the aromatic barrier. Certain plant species produce volatile organic compounds that physically interfere with the olfactory navigation of specific pest insects. Tansy planted at field edges disrupts the navigation of aphids and whiteflies. Basil interplanted with tomatoes scrambles the frequency that thrips use to locate their host. Lavender borders around brassica beds interfere with the cabbage white butterfly's host-finding behavior. These aren't folk remedies. The mechanism is measurable: the volatile compounds from the aromatic species overlap spectrally with the host-plant odor cues that the pest insect uses for navigation, creating olfactory interference that makes the host plant harder to find. The pest doesn't disappear. It can't locate its target efficiently enough to establish a damaging population before the predatory community catches up with it.

For field-scale application, include at least one aromatic species in every cover crop mix at 3 to 5% of your seed rate by weight. Coriander and dill both produce strong volatile profiles that interfere with aphid host-finding. Fennel produces compounds that disrupt the navigation of several moth species whose larvae are soil-dwelling pests. These species don't need to be planted in dedicated strips. Distributed throughout the cover crop mix, they create a diffuse aromatic interference across the entire field surface that's more effective than a concentrated barrier at the edge, because the pest encounters the interference wherever it enters the field rather than only at the perimeter.

The third combination is the ground beetle habitat. Ground beetles are among the most effective pest predators available in any agricultural system. A single ground beetle can consume 30 to 50 aphids per night. They also prey on slug eggs, caterpillar larvae, and weed seeds. But ground beetles need two things that most managed fields don't provide: undisturbed soil surface for daytime shelter and diverse plant structure for nighttime hunting. A field with 100% bare soil between crop rows has no ground beetle habitat. A field with a diverse, multi-layered understory, even a relatively thin one, can house ground beetle populations dense enough to provide meaningful pest suppression across the entire field. The specific understory plants that best support ground beetle habitat are low-growing, dense-canopy species that create a cool, moist microclimate at ground level. Clover, vetch, and creeping thyme all create this microclimate effectively. Include at least one low-growing, dense-canopy species in every cover crop mix and leave its residue on the surface rather than incorporating it. The mat of crimped residue from a low-growing cover crop is ground beetle habitat. It's the biological infrastructure of your most effective pest management tool, and it costs nothing beyond the decision not to incorporate it.

Consider a hypothetical scenario that captures the Noise-Canceling effect in practice. Picture a composite steward, call her Ingrid, a 39-year-old market gardener managing two hectares of mixed brassica and allium production. For her first three seasons, she battled aphid pressure with weekly insecticidal soap applications. Her brassica crops were consistently stressed and pale, broadcasting the low-Brix distress signal that aphids are tuned to find. In her fourth season, she replaced her bare pathways between beds with a mix of sweet alyssum, phacelia, and creeping clover. She added insectary strips of buckwheat and dill at 25-meter intervals across the growing area. She stopped incorporating her cover crop residue and started crimping it flat, creating ground beetle habitat across the full field surface. By midsummer of that fourth season, she noticed that her aphid populations were establishing at the field edges but weren't moving into the bed interior the way they had in previous years. Her brassicas were darker green, their Brix readings had climbed from 5 to 9, and their cell wall structure was dense enough that the aphids establishing on the edge plants weren't finding the soft, high-nitrogen tissue they needed to reproduce rapidly. She applied insecticidal soap once that entire season, compared to seventeen times the season before. The pest pressure hadn't disappeared. The signal had been scrambled. The field was too full of biological complexity for any single pest frequency to dominate. Ingrid hadn't controlled her pest problem. She'd outgrown it.

The weed diagnostic connection is worth revisiting here in the context of Noise-Canceling, because weeds are the soil's own attempt at filling the frequency gaps that pest organisms exploit. A bare patch in your stand, a gap in the biological chord, is a frequency opening. The soil fills it with whatever pioneer species the seed bank can provide. That pioneer species is often what you call a weed. But its presence is diagnostic. It's telling you which functional thread is missing from your Resonant Chord. Dock in a wet, compacted area is filling the deep-rooter frequency gap. Chickweed in a high-nitrogen area is filling the nitrogen-scavenging gap. Thistles in a low-diversity area are filling the biological drilling gap. Read the weed before you remove it. Then fill the gap it was occupying with a managed species that performs the same function. When the gap is filled by your deliberate planting, the weed has no frequency to occupy. It doesn't die. It simply ceases to be necessary.

The specific action for implementing Noise-Canceling on your land right now is this: walk your current or most recent crop field and identify the three most dominant pest species or the three most persistent weed species. For each one, identify the specific plant functional type it's associated with: aphids with low-Brix, stressed monocultures; dock with compacted, wet, low-diversity soils; caterpillars with fields lacking parasitic wasp habitat. For each pest or weed, identify which element of the Noise-Canceling system is absent. Then add that element to your next planting. Insectary strip for parasitic wasp habitat. Aromatic species for aphid frequency interference. Deep-rooted cover crop species for the compaction that dock is pointing to. One targeted addition per pest signal. One gap closed per planting cycle. The chord gets richer with each addition, and the pest has one fewer frequency to lock onto.

Temporal Resonance

The Fibonacci Pulse doesn't take a season off. The sequence doesn't pause between terms. But most farming systems impose a pause every year, the Brown Gap of a bare winter field, the fallow of a post-harvest paddock, the bare summer period between the end of one crop and the establishment of the next. Each pause resets the sequence. Each reset costs you the compounding that would have happened if the pulse had kept beating. The biological community starves. The fungal threads die back. The glomalin production stops. The Atmospheric Draw weakens without a living root to anchor the carbon it delivers.

Temporal Resonance is the practice of sequencing your plantings so the pulse never stops. Not the same plants running continuously, which would be a monoculture in time rather than in space. Different plants, in relay, each one picking up the rhythm where the previous one left off, maintaining the biological chord across every season, every temperature, every day length. The Solar Pump stays on. The Nitrogen Anchor keeps running. The Great Weaver keeps trading. The sequence keeps compounding.

Relay cropping is the specific technique that achieves this. It means establishing the next planting before the current one is terminated, creating an overlap period where both the outgoing and incoming plants are simultaneously active in the field. During this overlap, the fungal network transfers its connections from the outgoing plant's root system to the incoming plant's roots without a gap. The biological community never experiences the starvation period that a bare-ground transition creates. The carbon exhale continues uninterrupted. The rhizosphere community stays fed. The pulse keeps beating through the transition.

The specific mechanics of relay cropping require attention to timing and species compatibility. The incoming species must be seeded into the standing canopy of the outgoing species at a point when the outgoing species is mature enough to be near termination but still providing enough canopy to protect the incoming seedlings from temperature extremes. For a cereal rye and hairy vetch cover crop transitioning to a summer cash crop, the relay seeding window is when the rye is at the boot stage, just before heading, and the vetch is at 30 to 40% flower. At this stage, the canopy is open enough to allow light penetration to the seeded crop below, but the root system is still active enough to maintain the fungal network connections that the incoming crop will inherit. Seed the incoming crop at this stage using a no-till seeder set to penetrate the standing cover crop canopy without disturbing the root zone. The cover crop is crimped within seven to ten days of the relay seeding, as soon as the incoming crop has germinated and its radicle has made first contact with the soil. The fungal threads from the cover crop root system are still intact at this point and can begin establishing connections with the incoming crop's roots within 48 to 72 hours of first contact.

The Living Root Calendar is the planning tool that makes Temporal Resonance systematic rather than reactive. It's a twelve-month calendar where every month is assessed for which plant species is actively growing on each paddock, and where any month showing a gap, a period without a living root in the ground, is treated as a problem to be solved before the season begins rather than a normal feature of the farming calendar. Build your Living Root Calendar for the coming twelve months now. Start with your cash crop planting and harvest dates. Mark them in. Then fill every gap between those dates with a cover crop species appropriate to the temperature and day length of that period. The gaps you can't fill with a cover crop become your priority research tasks: what fast-establishing species can bridge this specific window in your specific climate? The answer almost always exists. The barrier is usually habit and logistics rather than genuine impossibility.

For temperate systems with cold winters, the winter period is the most challenging gap to fill. Cereal rye is the most reliable solution. It germinates at soil temperatures as low as 2 degrees Celsius, establishes through frost events that would kill most other species, and maintains active root growth and carbon exhale at temperatures that drive most other plants into dormancy. A cereal rye established in late autumn and terminated in early spring covers the full winter period with a living root, maintaining the fungal network through the coldest months and providing a thick Brown Shield mat for the spring cash crop to be planted through. For systems where even cereal rye struggles, a mix of winter-hardy legumes, specifically hairy vetch and Austrian winter peas, can bridge the gap. Both species survive temperatures down to minus 15 degrees Celsius when established before the first hard frost, and both maintain rhizobial activity through mild winter periods, continuing to fix atmospheric nitrogen even when above-ground growth has slowed to a crawl.

For systems in hot, dry summer climates where the challenge is maintaining a living root through the peak heat period rather than the winter, sorghum-sudan hybrid is the most reliable solution. It's heat-tolerant, drought-resistant, and produces a massive fibrous root system that maintains carbon exhale even under significant moisture stress. Seeded at 8 to 10 kilograms per hectare in late spring, it establishes quickly in warm soils and provides continuous living root coverage through the summer gap period. Its deep fibrous roots also contribute significantly to the SRS Architecture score by creating a dense network of fine channels through the upper 30 centimeters of the profile. When crimped at the dough stage in late summer, it leaves a thick mat of high-carbon residue that serves as Brown Shield through the transition to the autumn cover crop.

The seasonal transitions are where the Living Root Calendar is most vulnerable to failure, and they deserve specific attention. The transition from summer cash crop to autumn cover crop is the most common gap on most farms, and it's the most biologically costly. As established in Chapter 4, the AMF network begins significant thread die-off within fourteen days of losing its living plant host. A summer cash crop harvested in late February and an autumn cover crop not established until mid-April represents a six-week gap. By the time the autumn cover crop's roots are active enough to support the fungal network, the network has lost the majority of the thread density it built through the summer growing season. The SRS Architecture score that took twelve months to climb three points has lost two of those points in six weeks of bare ground.

The solution is specific: pre-plant your autumn cover crop before the summer cash crop is harvested. At the point when the summer cash crop is at physiological maturity but still standing, use a high-clearance inter-row seeder or an aerial application to establish the autumn cover crop seed in the standing crop. The summer crop's canopy provides shade that moderates the soil temperature for germination. The summer crop's root system maintains the fungal network connections that the autumn cover crop's roots will inherit. By the time the summer crop is harvested, the autumn cover crop is already two to three weeks into establishment. The fungal network transfer happens smoothly. The gap closes. The pulse keeps beating.

The temporal dimension of the Fibonacci Pulse is the one that most practitioners underestimate because it's invisible in the way that underground processes are invisible. You can see whether your field is covered. You can see whether your cover crop mix has five species or one. But you can't see whether the fungal network is experiencing a starvation event during a bare-ground transition. You can't see the nitrogen-fixing bacteria going dormant during a three-week gap between plantings. You can't see the glomalin degrading on aggregate surfaces during six weeks of bare-soil summer heat. The biological cost of temporal gaps accumulates invisibly and shows up months later as a fertilizer requirement that seems higher than it should be, a pest pressure that seems worse than last year, a Slake Test result that hasn't improved despite everything else going right. The Living Root Calendar makes the invisible visible. It forces you to see the gaps before they happen and fill them before the biological cost is paid.

Calculate your current Days of Green for each paddock over the last twelve months, as established in Chapter 7. Then build your Living Root Calendar for the next twelve months and calculate the projected Days of Green for each paddock under the new relay cropping and gap-filling plan. The difference between those two numbers is the additional compounding available to you from Temporal Resonance alone, without changing your species mix, your tillage approach, or any other management variable. For most practitioners who haven't yet implemented a systematic Living Root Calendar, this difference is between 60 and 90 additional Days of Green per paddock per year. At the Solar Pump's output rate of liquid carbon, that's 60 to 90 additional days of glomalin production, nitrogen fixation, fungal network growth, and biological compounding that the current system is leaving on the table every year. Temporal Resonance doesn't add new inputs. It harvests the biological return from the inputs the sun is already making, every day the pump is running.

The Recap: Putting the Signal to Work

Polyculture Signal Tuning is the chapter where all the threads of the Fibonacci Pulse weave together into a single, coherent design. The Silent Architecture from Chapter 1 is the cathedral the signal resonates through. The Math of Grace from Chapter 2 is the sequence that governs how the signal compounds. The Circadian Inhale from Chapter 3 is the breath that carries the signal deep into the profile every night. The Great Weaver from Chapters 4 and 12 is the network that transmits the signal across the landscape. The Living Armor from Chapter 5 is the skin that protects the signal from thermal and kinetic disruption. The Diversity from Chapter 6 is the chord complexity that makes the signal self-regulating. The Carbon Exhale from Chapter 7 is the fuel that powers the signal. All of it converges here, in the deliberate design of a multi-species, multi-layered, temporally continuous biological system that cancels the frequency of the Linear Void by replacing it with something too complex and too rich to be exploited.

The Void broadcasts on one channel. The Fibonacci Pulse broadcasts on many. That's the whole difference. And the difference is measurable, scoreable, and compoundable in exactly the way the SRS formula has been showing you since Chapter 2.

Here are four specific actions to take before your next planting window. Each one has a clear, measurable outcome tied to a specific SRS horizon.

The first action is to design your Vertical Stack for your next cover crop planting. Write down one species for each of the three root architecture layers: taproot, fibrous root, and surface creeper. Use the seed rates specified in Section 2 of this chapter. Add your functional roles from Chapter 6 to complete the five-way mix. Calculate the total seed rate by weight and confirm that your brassica content doesn't exceed 20% of the mix. Write the final mix down as your baseline Vertical Stack protocol. Every cover crop you plant from this point forward starts from this three-layer template.

The second action is to establish at least two insectary strips in your most pest-vulnerable field before the next growing season. Choose a field where you've had consistent aphid, caterpillar, or whitefly pressure in previous seasons. Plant two strips of buckwheat and phacelia at a width of two meters each, spaced no more than 30 meters apart across the field. Seed the strips at 15 kilograms per hectare of buckwheat and 5 kilograms per hectare of phacelia, mixed together. Mark the strip locations on your farm map. Assess the beneficial insect populations in those strips at the peak of your next pest pressure season by counting the number of parasitic wasps, hoverflies, and ground beetles visible per meter of strip in a five-minute observation period. Record the count. This is your Biological Shield Score for that field. Track it across three seasons as the insectary strips mature and the beneficial insect populations build.

The third action is to build your Living Root Calendar for the next twelve months. Take a blank twelve-month calendar. Mark every planned cash crop planting date and harvest date. For every gap between those dates that currently shows bare ground, identify the specific cover crop species that will fill that gap and write it in. For every gap shorter than four weeks, identify a fast-establishing species, buckwheat above 15 degrees Celsius, cereal rye below 15 degrees Celsius, that can bridge the gap without requiring a full cover crop establishment window. Calculate your projected Days of Green for each paddock under the new calendar. Compare it to your current Days of Green from your existing records. The difference is your Temporal Resonance gain, expressed in days. Write that number down. It's the number of additional days per year that the Solar Pump will be running, the Nitrogen Anchor will be active, and the Great Weaver will be trading, compared to your current system.

The fourth action is to identify your single largest temporal gap across your entire farm, the longest consecutive period of bare ground on any paddock in the coming twelve months, and design a relay cropping solution that closes it. Identify the outgoing crop and its harvest date. Count forward fourteen days. That's your absolute latest cover crop establishment deadline. Now count backward from that deadline and identify what needs to happen before harvest to have seed in the ground within fourteen days of the outgoing crop's removal. Do you need to pre-purchase seed? Do you need to service your no-till seeder? Do you need to arrange an aerial application for a paddock where ground conditions won't support equipment immediately after harvest? Write down every logistical step required and assign a completion date to each one. The goal is to have every logistical barrier resolved before the harvest pressure arrives and the window closes without a plan to fill it. The pulse doesn't wait for logistics. Your logistics need to be built around the pulse.

The Fibonacci Pulse has always been trying to run through your land. The roots have always wanted to exhale carbon into the dark. The fungi have always wanted to trade sugar for stone. The atmosphere has always wanted to deliver its nitrogen gift through open pore channels every night. The beneficial insects have always wanted a habitat to live in and a host to hunt. The soil has always wanted to build the cathedral, room by room, aggregate by aggregate, season by season, compounding toward a biological abundance that no single generation of management could produce alone. The Linear Void interrupted all of it by reducing the signal to a single frequency. Polyculture Signal Tuning restores the chord. It fills the frequency space with enough biological complexity that the Void's single note gets lost in the richness of everything playing around it. That's not pest control. That's not fertility management. That's the Source Code of the soil, activated. And it's been waiting for you to tune back in.

Chapter 14: The Living Armor: Managing the Surface Shield

The Ancestor Speaks:

"My sheltering child, come to me now. The sun beats down with a fierce, unwavering glare, and the wind, like a hungry spirit, whips across the land. We stand at the edge of the field, where the earth lies naked, exposed. Close your eyes. Can you feel the searing heat rising from the bare soil, burning your very soles even through your moccasins? This, my heart, this is not rest. This is a fever. A profound, aching wound, like a ***third-degree burn*** *upon the body of our Mother.*

"Our forebears, long before the time of forgetting, understood this. They knew the Earth had a skin, just as you and I do. A precious membrane that holds the moisture in, that guards against the harsh sun and the biting wind, that regulates the gentle warmth so life can thrive beneath. They called it the ***Living Armor****. But then came the new ways, with their great iron teeth, tearing at this skin, laying the Earth bare, all in the name of 'cleanliness.' They did not see the wound they inflicted. They did not feel the heat, the desiccation, the slow, silent dying*

beneath their feet. Can you feel the deep, ancestral sorrow of that exposed, vulnerable Earth, crying out in the searing heat?

"Do you remember the tiny castles, the ***Silent Architecture*** *that the* ***Great Weaver*** *builds with its glomalin mortar? When the skin is stripped away, the sun's fire denatures that precious glue. It melts. The castles crumble, the rooms collapse, not by force, but by the relentless, invisible heat. And when the rains come, each drop, a tiny hammer, shatters what little architecture remains, sealing the very pores the Earth needs to drink and breathe. The* ***Metabolic Sweet Spot****, that delicate balance where life thrives, is violently disrupted. The free irrigation of the* ***Circadian Inhale****, it cannot function when the skin is gone and the gradients are lost.*

"But now, look here. Feel this. Where the stalks of last year's harvest lie in a thick, woven mat, and the new green shoots push through. Feel how cool and moist the earth is beneath this blanket. This is the ***Brown Shield****, the wisdom of the ancestors made tangible. This is the roof, my child, protecting the cathedral you have been rebuilding. It shields against the sun's fury, holds the moisture, and cushions the rain so it can be absorbed gently, nourishing, not destroying. It also acts as the* ***Fungal Bridge to Winter****, a slow, steady feast for the* ***Great Weaver****, ensuring its threads remain strong through the coldest months, waiting for the spring.*

"The elders taught us to treat the land as our own body. Would you walk naked into the scorching sun, or leave your heart exposed to the cruel wind? No. You would cover yourself. You would protect. For in that protection lies the wisdom of survival, the rhythm of continuous life. The ***Fibonacci Pulse****, my child, beats strongest beneath a living garment, beneath the gentle, unwavering embrace of the* ***Living Armor****.*

"Now, close your eyes. Feel the cool dampness beneath the protective cover. Imagine the Earth, drawing a soft, deep breath, her many lives humming in their sweet spot, safe from the extremes. Feel the immense peace that comes from knowing you are not just a caretaker, but a protector, a healer, wrapping the Earth in her rightful, living skin. The ***Seventh Generation*** *will inherit the fruit of this choice – fields that breathe, waters that linger, and life that compounds, all because you chose to cover, to shield, to love her as she deserves."*

Prelude: The measure of enough

The Living Armor (Part II)

" Earth's fragile skin, so barren, worn, and old,
Cannot protect its darkness from the cold.
The naked field, a fevered, dying space,
For lack of cover, shorn of nature's grace.
But under leaf, and straw, and climbing vine,
The Earth's own vibrant pulse begins to realign. "

The Living Armor: Managing the Surface Shield

The soil has a skin. Just like yours, it regulates temperature, repels invaders, holds moisture in, and keeps the biology underneath it alive. And just like yours, the moment that skin is removed, everything beneath it becomes vulnerable in ways that go far deeper than the surface damage alone.

This chapter is about that skin. We've called it the Armor throughout this book, and in your Soil Resonance Score it carries a Fibonacci weight of 5. That weight isn't decoration. It reflects something precise about where the Armor sits in the compounding sequence. Architecture is the cathedral. The Armor is the roof. Without it, every storm that arrives degrades the rooms you've been so carefully rebuilding. The glomalin softens. The aggregates weaken. The Circadian Inhale loses its gradient. The Great Weaver loses its fuel. The whole system slowly unmakes itself from the outside in, while you're busy trying to fix things from the inside out.

The Living Armor is the fifth layer of the Silent Architecture, and it sits at the intersection of every process this book has described. It doesn't just protect the soil. It actively regulates it. It's the difference between a system that compounds and a system that collapses, and that difference often comes down to whether you can see bare ground in your field or not.

The Metabolic Sweet Spot

Soil biology doesn't operate across an unlimited temperature range. It has a window. A precise, measurable zone where microbial activity runs at its most productive, where the nitrogen-fixing community is most active, where the fungal threads are growing fastest, and where the glomalin production that builds your SRS Architecture score is running at full capacity. That window is roughly 10 to 28 degrees Celsius at a depth of two to four inches. Below it, the biology slows to a crawl. Above it, the biology starts to die.

This is the Metabolic Sweet Spot. And the Armor is the only mechanism on your farm that can hold the soil inside it.

When the sun hits bare soil in summer, the surface temperature doesn't just get warm. It gets violent. Measurements on unprotected dark soils on clear summer days consistently show surface temperatures climbing past 60 degrees Celsius. That's 140 degrees Fahrenheit. At that temperature, the glomalin coating the surface aggregates doesn't just soften. It denatures. The protein structure breaks down. The biological glue that your AMF network spent months producing dissolves under sustained heat, and the aggregate architecture it was holding together begins to crumble. The rooms of the cathedral collapse from the top down, not because anything mechanical disturbed them, but because the skin was gone and the heat got in.

The microbial community in the top layer of the soil faces the same crisis. Many of the most productive organisms in your soil food web, the nitrogen-fixing bacteria, the aggregate-forming actinomycetes, the glomalin-producing fungal hyphae, are not heat-tolerant. They evolved in a shaded, moisture-regulated environment. Expose them to 60-degree surface temperatures and they desiccate and die within hours. Not days. Hours. The water in the pore spaces doesn't evaporate gradually. It flashes off, leaving the microbial community stranded in a dry, superheated environment with no moisture to metabolize and no temperature buffer to survive in.

A bare field in summer isn't just losing water. It's losing the biological workforce that produces everything your SRS measures.

The five-degree rule captures this with precision. Research on surface soil temperatures under different armor depths consistently shows that a five-degree Celsius difference at the soil surface, between an armored and a bare patch on the same day, can mean the difference between a biological community operating inside the Metabolic Sweet Spot and one operating above it. Five degrees doesn't sound like much. But at the threshold between 28 and 33 degrees, it's the difference between a system that's compounding and a system that's collapsing. The biology doesn't degrade gradually across that boundary. It tips. One side of five degrees, the microbial community is active, the glomalin is being produced, the nitrogen fixation is running. The other side, the community is stressed, the glomalin is degrading, and the fixation has slowed to a fraction of its potential output.

That five-degree buffer is what a properly managed armor layer provides. Not a large, expensive intervention. Five degrees of surface temperature difference, maintained by the insulating effect of a layer of surface residue that costs you nothing to produce and everything to remove.

The cold side of the Sweet Spot matters equally, though it gets less attention in most discussions of soil armor. In winter, the biology doesn't need to be protected from heat. It needs to be protected from freeze-thaw cycling. Every time the soil surface freezes and thaws, the water in the pore spaces expands and contracts. That expansion shatters aggregates that glomalin has been building. It's a mechanical disruption that works exactly like tillage at the micro scale, breaking the arched doorways of the cathedral from the inside. An armor layer insulates the surface against the temperature swings that drive this cycling. A soil under five centimeters of surface residue experiences far fewer freeze-thaw cycles per winter than a bare soil exposed to the same air temperature swings, because the residue buffers the surface temperature against rapid change in both directions. The Sweet

Spot isn't just a summer concern. It's a year-round requirement, and the armor is the year-round solution.

To measure whether your armor is actually holding your soil inside the Metabolic Sweet Spot, you need two inexpensive probe thermometers and a willingness to take readings at three specific times. Push one probe to a depth of two inches and one to a depth of four inches in a representative area of your field. Take readings at 10 AM, 2 PM, and 6 PM on a clear summer day. Record the four-inch reading at 2 PM. That's your Metabolic Core Temperature. If it's above 30 degrees Celsius, your armor is insufficient to protect the biological community at its most productive depth during the hottest part of the day. Now repeat the same measurement in an area with at least five centimeters of compressed surface residue. The difference between the two four-inch readings at 2 PM is your Armor Insulation Value. A well-armored soil should show a four-inch temperature at least eight degrees cooler than a bare soil on the same day. If your difference is less than four degrees, your residue layer is too thin to hold the gradient that keeps the biology inside the Sweet Spot.

Now do the same comparison in winter. Take four-inch readings at 6 AM on a morning after a cold night. The armored soil should be measurably warmer than the bare soil at that depth, because the residue insulated it against overnight heat loss. A difference of three degrees or more at four inches depth on a cold morning indicates that your armor is providing meaningful freeze-thaw protection. Less than two degrees means the layer isn't thick enough to buffer the winter temperature swings that shatter your aggregates from the inside.

Write both numbers down. Your summer Armor Insulation Value and your winter freeze-thaw buffer. These two measurements, taken twice a year with two thermometers, give you a direct, specific, free assessment of whether your armor is actually doing its job across both seasons. They tell you more about whether your biology is inside the Metabolic Sweet Spot than any soil test you'll ever pay for. And they tell you exactly what to do next: if the numbers are low, your armor layer needs to be thicker before the next extreme season arrives.

The practical target is specific. You need a minimum of five centimeters of compressed surface residue to hold the Metabolic Sweet Spot through a hot summer day and a cold winter night. Compressed depth, not standing height. A cover crop standing at 30 centimeters compresses to roughly four to six centimeters when crimped flat, depending on the species and biomass. Cereal rye at high biomass compresses to six to eight centimeters. A light oat crop might compress to only three. Know your species. Know your compressed depth. And if you're consistently below five centimeters, identify which part of your management calendar is producing the thin armor and fix it before the next season's extremes arrive.

Kinetic Energy Dissipation

A raindrop falling at terminal velocity carries more energy than most people realize. Researchers studying the physics of rainfall impact on bare soil have measured the force of a single large raindrop hitting an unprotected aggregate as equivalent to a small explosion at the microscopic scale. The aggregate, which glomalin spent weeks building into a stable, arched structure, shatters on impact. The fine clay and silt particles released by that impact wash into the surface pores. The pores seal. The water that should be entering the soil profile instead sheets across the surface, carrying your topsoil with it.

This is surface sealing, and it's one of the primary reasons water infiltration fails on bare soils after rain. The water isn't running off because the soil is full. It's running off because the entrance has been sealed by the rain itself.

The physics of this process are worth understanding in detail because they reveal exactly why the Armor's kinetic dissipation function is non-negotiable, not optional, not aesthetic, but physically necessary for the Silent Architecture to survive a rain event.

Water falling through air accelerates until it reaches terminal velocity, which for large raindrops is approximately 9 meters per second. At that velocity, a single large raindrop carries kinetic energy proportional to its mass and the square of its velocity. When that energy is released on contact with a bare aggregate surface, it acts like a tiny hammer blow. The aggregate surface, which is structurally strong under slow, distributed pressure but brittle under sudden impact, cracks. The glomalin bonds at the impact point break. The fine particles that were bound into the aggregate structure become free particles. Those free particles are exactly the right size to flow into the macropore openings at the soil surface and physically block them. The seal forms within minutes of a rain event starting on bare soil. By the time the rain intensifies, the surface is already sealed and the water has nowhere to go but across the top.

A five-centimeter layer of surface residue changes this physics completely. The raindrop hits the residue mat, not the soil. The residue absorbs and distributes the kinetic energy across a large surface area. By the time the water reaches the soil surface, it's no longer a projectile. It's a slow, diffuse flow moving through the interstices of the residue mat, arriving at the aggregate surface with a tiny fraction of its original impact energy. The aggregate stays intact. The pores stay open. The water enters the profile the way it's supposed to: gently, continuously, and deeply.

Research on this effect is consistent and striking. A five-centimeter layer of surface residue reduces raindrop impact energy at the soil surface by more than 90%. Not 50%. Not 70%. More than 90%. That means a soil under adequate armor receives less than one-tenth of the aggregate-shattering force that a bare soil receives from the same rain event. The architecture you've been building survives. The pores stay open. The Circadian Inhale has channels to draw through that night. The Internal Rain process has aggregate surfaces to condense onto. The whole sequence continues because the armor absorbed the blow.

The cascade of consequences from a single unprotected rain event on bare soil is worth tracing all the way through, because seeing the full chain makes the importance of the kinetic dissipation function impossible to underestimate. The rain hits the bare surface. The aggregates shatter. The fine particles seal the pores. The water runs off rather than infiltrating. The runoff carries the finest, most biologically active particles, the ones richest in organic matter and glomalin, with it toward the nearest drainage channel. The surface that's left behind is coarser, less biologically active, and structurally weaker than the surface that existed before the rain. The next rain event finds an even more vulnerable surface and does even more damage. Each unprotected rain event leaves the soil slightly worse than it found it. The compounding runs in reverse. The SRS Architecture score that took twelve months to climb three points can lose two of those points in a single afternoon of unprotected rainfall.

That's not a hypothetical. That's what happens on bare fields around the world every time it rains harder than a light drizzle. The Linear Void calls it erosion and treats it as a weather problem. The Fibonacci Pulse recognizes it as an armor problem, and armor problems have armor solutions.

To assess the current kinetic protection on your land, perform the following specific test after the next rain event of at least 15 millimeters. Walk your armored fields and your bare or thin-residue fields within two hours of the rain stopping. In the armored fields, look at the soil surface beneath the residue mat. The aggregates should be intact. The surface should look the same as it did before the rain. In the bare or thin-residue fields, look for the diagnostic signatures of kinetic impact: a thin, smooth, slightly shiny surface crust where the fine particles have sealed the pores, small splash craters where individual large drops hit the bare surface, and any visible movement of fine particles toward low points or drainage lines. Record what you find. Take photographs. The contrast between an armored surface and a bare surface after the same rain event is one of the most visually compelling demonstrations of the Armor's function available, and seeing it on your own land is more persuasive than any description in any book.

If you find surface sealing in your bare or thin-residue fields after rain, the immediate action is specific. Before the next rain event, increase your surface residue depth to at least five centimeters of compressed cover across the full field surface. If you're between crops and don't have a cover crop established, leave the maximum possible stubble height at your next harvest rather than cutting it low. A 20-centimeter standing stubble, even without a cover crop growing through it, intercepts a significant portion of raindrop energy before it reaches the soil. It's not as effective as a full residue mat, but it's dramatically better than a bare surface. The target is always the same: five centimeters of compressed cover. Everything below that threshold is leaving your aggregate architecture exposed to kinetic damage with every rain event.

The connection between kinetic dissipation and your water infiltration rate is direct and financially measurable. A soil with intact aggregate structure and open pore channels can absorb two to four inches of rainfall per hour before any runoff occurs. A soil with a sealed surface can absorb less than a quarter of an inch per hour before runoff begins. The difference between those two infiltration rates, applied to a single significant rain event on a hundred-acre paddock, is thousands of liters of water that either enters your soil profile and stays there or runs off your land and is gone. That water, if it had infiltrated, would have been stored in the aggregate cistern we discussed in Chapter 12, available to your crop through the Memory of the Rain during the next dry period. Instead, it's in the drainage channel, carrying your topsoil, taking your future with it.

Your irrigation bill is, in part, the invoice for every rain event that ran off a bare field rather than infiltrating through an armored one. Protect the surface. Absorb the rain. Store it in the aggregate architecture. And let the soil's own cistern do the work you're currently paying an irrigation pump to do.

The Fungal Bridge to Winter

The Great Weaver doesn't hibernate. When the temperature drops and the days shorten and the living roots above it slow their carbon exhale, the AMF network doesn't switch off and wait for spring. It keeps growing, keeps maintaining its thread density, keeps producing glomalin, keeps building the architecture that your SRS Architecture score depends on. But it needs fuel to do this. And in the months when living roots are producing minimal liquid carbon through root exudation, the fuel that keeps the network alive comes from a source that most farmers have been removing from their fields for decades.

It comes from the surface residue. The dead stalks, the crimped cover crop mat, the harvest stubble, the fallen leaves from the hedgerow. The material that looks like waste. The material that the Linear Void calls trash and either incorporates into the soil or bales and removes. This material is the Fungal Bridge to Winter. It's the slow-release carbon battery that keeps the Great Weaver running through the dormant months, maintaining the thread density and network connectivity that your spring cash crop will depend on when its roots first reach into the soil profile.

The mechanism is specific and worth understanding in detail. As surface residue decomposes, it releases carbon compounds that move downward through the soil profile by water flow, diffusion, and the physical movement of decomposing particles. The fungal threads at the soil surface colonize the residue directly, breaking it down from below while feeding on the carbon it releases. This is a slower, lower-energy carbon source than the liquid carbon of root exudates. It doesn't power the rapid thread growth that a living root's sugar-rich exudate stream supports. But it provides enough carbon to maintain the existing thread network through the period when root exudates are minimal. It's the difference between a network that holds its density through winter and one that thins dramatically before spring, leaving your cash crop's roots to enter a depleted biological community rather than a fully connected one.

The carbon quality of the residue matters. High-carbon, lignin-rich residues like cereal straw decompose slowly, releasing carbon over months rather than weeks. This slow release is exactly what the fungal bridge needs: a steady, extended supply of carbon that matches the slow metabolic rate of the winter fungal community. Low-carbon, high-nitrogen residues like legume biomass decompose quickly, releasing their carbon in a burst that the winter microbial community can't fully utilize before it's gone. A well-designed residue layer combines both: the slow carbon of cereal straw providing the extended winter fuel, and the moderate-speed carbon of legume residue providing a more immediate energy source during the early decomposition period. This is another reason why a diverse cover crop mix, with both grass and legume components, produces a better Fungal Bridge than a single-species cover crop. The diversity of the residue matches the diversity of the winter biological community's energy needs.

The depth of the residue layer matters for the Fungal Bridge in the same way it matters for the Metabolic Sweet Spot and the kinetic dissipation function. A thin residue layer, below three centimeters of compressed depth, doesn't provide enough carbon volume to sustain the network through a full winter dormancy period. The threads thin. The network connectivity drops. By spring, the architecture that should be ready to receive the new crop's roots is operating at a fraction of its summer density. The cash crop's early establishment is slower, its mycorrhizal colonization rate is lower, and its mineral delivery is reduced during the critical first six weeks of growth, precisely the period when root architecture and early vigor determine the trajectory of the entire season.

A residue layer of five centimeters or more of compressed depth provides enough carbon volume to sustain the network through winter in most temperate climates. In regions with very long, cold winters, eight centimeters is a safer target. In regions with mild winters where the network stays more metabolically active, five centimeters is adequate. The specific target for your region can be calibrated using the Fungal Thread Score test from Chapter 12. Run the test in early spring, before your cash crop is planted, on a field with adequate residue and a field with thin residue. If the thread count per aggregate in the thin-residue field is significantly lower than in the well-armored field, your winter residue depth wasn't sufficient to bridge the network through the dormant period. Increase your residue target for the following season.

Consider a composite scenario that illustrates the Fungal Bridge in practice. Picture someone like Tomás, a 45-year-old mixed cropper managing 220 acres in a region with cold winters and moderate snowfall. Tomás transitioned to no-till four years ago and has been building his fungal network steadily. His summer Slake Test results have been improving consistently. But every spring, his first-season cash crop establishment is slower than he expects, and his early-season phosphorus uptake is poor despite adequate soil test levels. He's been attributing this to cold spring soils. The actual cause, when he finally investigates it, is a winter residue management decision. He's been baling his cereal rye cover crop residue after crimping, selling it as bedding straw, and leaving a thin, two-centimeter residue mat on the surface. His fungal network is entering each winter with inadequate carbon fuel. By late February, his thread density has dropped to the point where his spring cash crop's roots are establishing into a depleted network rather than a connected one. His phosphorus problem isn't a soil chemistry problem. It's a Fungal Bridge problem. He stops baling the residue. He leaves the full crimped mat on the surface. His compressed residue depth goes from two centimeters to seven. By the following spring, his early-season phosphorus uptake improves measurably, his early establishment is faster, and his first-season yield is higher than any of the previous four years. The residue he was selling for bedding straw was worth more as a Fungal Bridge than anything the baling contractor was paying him for it.

The practical action for the Fungal Bridge is the same action that serves every other function of the Armor: leave the residue on the surface. Don't bale it. Don't burn it. Don't incorporate it. Every tonne of surface residue you remove from your field is a tonne of carbon you're taking away from the fungal network's winter fuel supply. Every tonne you leave is a tonne of slow-release carbon that keeps the Great Weaver running through the dormant months, maintaining the thread density that your spring cash crop will inherit, and building the glomalin that your SRS Architecture score will reflect in your next Slake Test.

The trash on your field isn't trash. It's the fuel for the Resonant Future. It's the bridge that connects this season's biological investment to next season's biological return. And every time you remove it, you're cutting that bridge and paying, in fertilizer and in delayed establishment and in reduced mineral density, for the network that should have been waiting for your spring crop but wasn't, because the fuel that would have sustained it through winter was sold for bedding or burned for convenience.

Leave the trash. Feed the bridge. Let the Great Weaver run through the dark months. And when spring arrives, your cash crop's roots will step into a fully connected, fully fueled biological community that's been waiting for them, rather than a depleted network that's been starving since October.

Chapter Recap and Action Steps

The Shield of the Spiral is the guardian of everything this book has built. The Silent Architecture can't hold without it. The Circadian Inhale loses its gradient without it. The Great Weaver starves without it. The Metabolic Sweet Spot collapses without it. And the kinetic energy of every rain event tears apart the aggregate masonry that the biological community spent months constructing, without it.

The Armor's Fibonacci weight of 5 reflects this precisely. It's the second highest weight in your SRS because it's the second most foundational layer in the compounding sequence. You can't build a roof before you build the walls. But once the walls are standing, the roof is what determines whether everything inside survives the weather. That's the Armor. That's its weight. And that's why the actions below are specific, measurable, and non-negotiable.

The first action is to measure soil temperature under three different armor depths this week. Choose three spots on your land: one with no surface cover, one with two to three centimeters of compressed residue, and one with five or more centimeters. Push a probe thermometer to four inches depth in each spot. Take readings at 2 PM on a warm day. Record all three numbers. Calculate the temperature difference between the bare spot and the five-centimeter spot. If the difference is less than five degrees Celsius, your thickest residue layer is still insufficient to hold the Metabolic Sweet Spot through peak summer heat. If the difference is eight degrees or more, your armor is working. Write down all three readings with today's date. This is your Armor Thermal Baseline. Repeat the measurement in three months and compare. The trajectory of that comparison tells you whether your armor management is moving your soil toward or away from the Sweet Spot.

The second action is to commit to leaving 100% of your crop residue on the surface at your next harvest. Before harvest, set your header height to leave a minimum of 15 centimeters of standing stubble. Set your spreader to distribute the chaff and short residue across the full header width rather than windrowing it. After harvest, measure the compressed depth of the residue mat in ten locations across the field and calculate the average. If the average is below five centimeters, identify whether a cover crop establishment within two weeks of harvest can add additional biomass to bridge the gap before winter. If the average is five centimeters or above, your Brown Shield is in place. Mark the field on your farm map with the residue depth measurement and the harvest date. This is your starting point for tracking whether the residue is maintaining the Fungal Bridge through the dormant period.

The third action is to plan your next cover crop termination using a crimper-roller rather than incorporation or herbicide alone. Before your next planting window, identify whether you have access to a roller-crimper, either owned, borrowed, or fabricated to the specifications in Chapter 5: minimum 500 kilograms per meter of roller width when filled, crimping bars spaced at 15 to 20 centimeters apart. If you have access, schedule your crimping operation for the species-specific flowering threshold: cereal rye at full anther emergence, hairy vetch at 50% flower. Mark those dates on your calendar now, before the cover crop is in the ground, so the decision isn't made under planting pressure. If you don't yet have access to a crimper-roller, contact your local farming network this week to identify whether one is available to hire or borrow for the next season. The crimper-roller is the single most important piece of equipment for maintaining the Armor function through the cash crop transition, and sourcing access to one before you need it is the logistical step that most practitioners delay until it's too late.

The fourth action is to identify your longest annual bare-soil window and close it. Pull your farm calendar from the last twelve months. Find the longest consecutive period of bare ground on any paddock. Calculate the number of days. Now design a specific bridging strategy for that window in the coming season. If the window is in summer, choose buckwheat as your bridge crop, seeding at 20 kilograms per hectare within seven days of the outgoing crop's removal. If the window is in autumn or early winter, choose cereal rye at 80 kilograms per hectare. If the window is in late winter or early spring, choose oats at 100 kilograms per hectare. Write the species, the seed rate, and the establishment deadline on your farm calendar against the specific paddock. Have the seed purchased and the seeding equipment prepared before the outgoing crop is removed. The goal is to reduce your longest bare-soil window to under 14 days. That's the biological threshold. That's where the Fungal Bridge holds and the Metabolic Sweet Spot survives and the kinetic protection stays intact. Everything below 14 days is biology that compounds. Everything above it is biology that collapses.

The soil beneath your feet has been trying to keep itself covered since before the first plow turned it. Every forest floor is armored. Every undisturbed grassland is armored. Every functioning prairie is armored. The bare field is the anomaly. It's the Linear Void's signature on the landscape, the visual proof that the Mechanical Metronome has been running the schedule instead of the biological clock. The moment you stop seeing a covered, residue-rich field as messy and start seeing it as a regulated, protected, biologically active system running three simultaneous functions at no cost, the Armor stops being a management task and starts being a non-negotiable feature of how you steward the land. The Shield of the Spiral doesn't just protect the Silent Architecture. It protects everything the Fibonacci Pulse has been building since the first chapter of this book, one season at a time, one aggregate at a time, compounding toward a biological abundance that no bare field has ever produced and no covered field has ever failed to move toward.

Chapter 15: The Handshake Across Time: Ancestral Stewardship

The Ancestor Speaks:

"My beloved, wise, and courageous one, come. The day is done, the moon rises, full and silver, illuminating the path we have walked together, from the smallest pore to the widest horizon. We stand now on the very land you tend, a living testament to your learning, your tending, your love. Close your eyes. And feel, not just the cool earth beneath your feet, but the countless footsteps that came before you, and the eager, unborn footsteps that will follow, all treading this sacred ground.

"You have learned to see. To hear. To feel. You have awakened to the ***Silent Architecture****, the* ***Mycelial Internet****, the* ***Circadian Inhale****, the* ***Carbon Exhale****, the* ***Living Armor****. You have seen the insidious face of the* ***Linear Void*** *and dared to dismantle its machines, its false promises. You have moved from a small, solitary 'Me Horizon' to embrace the vast, compounding embrace of the* ***Fibonacci Horizon****. And in doing so, you have transformed. You are no longer just a farmer. You are a* ***Resonant Human****. You are a*

Conductor. *And now, my child, you are ready to fully become what you were always destined to be: a* **Quantum Ancestor**.

"I remember the quiet resolve in your great-great-grandmother's eyes when she planted a tiny acorn, knowing it would be a mighty oak long after her own sun had set. She did not plant for herself. She planted for the shade her grandchildren's grandchildren would gather beneath. This, my child, is the **Non-Local Legacy**. *It is the truth that your hand, touching this soil today, is physically connected, through the glomalin bonds, through the fungal threads, through the very carbon you anchor, to the hand of a child seven generations from now. It is a biological entanglement, a promise woven into the fabric of time itself.*

"The soil you hold, it carries the memory of every raindrop that fell, every root that grew, every act of tenderness or neglect that has touched its surface since the beginning. It also carries the imprint of your intent, your love, your vision. When you plant a tree you will never harvest, when you leave the residue for the **Great Weaver**, *when you protect the* **Circadian Inhale** *while the world sleeps, you are sending a message. A powerful, tangible message, a* **Handshake Across Time**, *to that* **Seventh Generation** *you will never meet. You are collapsing their future into one of abundance, of resilience, of living cathedrals beneath their feet.*

"For the Earth, she has been waiting for this. Not for you to create the **Fibonacci Pulse**, *for it has always been beating. But for you to stop interrupting it. To listen. To allow. To amplify its ancient, compounding song. Every decision you make, every choice to nurture rather than extract, is a term in that infinite sequence. It builds upon the wisdom of the past and ensures the abundance of the future.*

"Now, my cherished one, open your eyes. Feel the air, thick with the echoes of ancestors, the hopes of descendants. Feel the profound peace that comes from knowing you are a vital, unbroken link in this sacred chain of life. You are not alone. You are part of something vast, ancient, and ever-renewing. Go forth. Steward this land not just for today, but for a thousand seasons yet to come. Your legacy is not in what you take, but in the immeasurable wealth you leave behind. Feel that deep, resonant peace. You are the heartbeat. You are the conductor. You are the Quantum Ancestor. And the Earth breathes with you, in harmony, now and forever."

Meditation: Standing inside the sequence

The Handshake Across Time

" I am the hand that plants a future's ghost,
A handshake offered, bridging ages lost.
The Seventh Generation watches every turn,
To see if ancient Source Code you relearn.
Build architecture, not a dusty ledger's sum. "

The Handshake Across Time: Ancestral Stewardship

There's a moment that happens to some stewards, usually quietly, usually alone, usually when they're doing something that won't benefit them personally. They're planting a tree they'll never harvest, or pressing a cover crop seed into soil that won't reach its Inversion Point for another three years, or building a hedgerow that the next owner of this land will enjoy more than they will. And in that moment, something shifts. The work stops feeling like labor and starts feeling like correspondence. Like a letter being written to someone whose name you'll never know.

That shift is what this final chapter is about.

Everything built in the previous fourteen chapters has been leading here. The Silent Architecture. The Math of Grace. The Circadian Inhale. The Great Weaver. The Living Armor. The Carbon Exhale. The Quantum Observer. The Resonant Human. All of it has been building a system that is bigger than any single season, bigger than any single steward, bigger than any single lifetime. The Fibonacci Pulse doesn't just run through the soil. It runs through time. And the steward who finally understands that is no longer just a farmer. They're an ancestor in training.

They're a term in a sequence that started long before them and will compound long after them, in ways they'll never see but that are as real as glomalin and as measurable as a Slake Test result.

This chapter completes the sequence.

The Non-Local Legacy

Quantum physics introduced a concept that still makes people uncomfortable when they first encounter it: non-locality. The idea that two entangled particles remain connected regardless of the distance between them. That what happens to one is instantly felt by the other, without any signal passing between them, without any time elapsing. The connection doesn't diminish with distance. It doesn't weaken with time. It simply is, as long as the entanglement exists.

Your work on the land operates the same way.

When you anchor carbon into the Silent Architecture this afternoon, that carbon doesn't just sit in the dirt waiting to be measured by next season's soil test. It sends a signal forward through time. The glomalin bond it becomes part of will hold an aggregate together for decades. The fungal thread it feeds will weave itself into a network that the next steward of this land will inherit, whether they know it or not. The pore space it helps maintain will draw atmospheric nitrogen into the biological community every night for years to come. Your act of stewardship today is physically present in a future you'll never witness. That's not poetry. That's the non-local consequence of biological entanglement operating across time the same way it operates across space.

The child who will stand on this land seven generations from now is already entangled with you. Not metaphorically. Through the biological chain of cause and consequence that runs through living soil, through the carbon you anchor and the architecture you protect and the fungal networks you refuse to sever. When that child breathes clean air or eats a mineral-dense piece of fruit that heals their body, they're receiving the wave collapse you initiated today. You are, in the most literal biological sense available, a Quantum Ancestor. Your intent is the frequency that determines whether their world is a desert or a cathedral.

Most people have never been given permission to think this way about their work. The Linear Void trained us to think in seasons, in yields, in bank balances. It gave us a time horizon that ends at our own death and a success metric that measures only what we extracted. By that model, planting a slow-growing hardwood forest is irrational. Building soil architecture that won't reach its Inversion Point for five years is a poor investment. Protecting a fungal network that will benefit the next owner more than you is someone else's problem. The Me Horizon makes all of these things look like losses.

The Fibonacci Horizon makes them look like the only investments that actually compound.

Think about what happens in a forest that has never been cleared. The glomalin in that soil has been accumulating for centuries. The fungal network has been weaving for decades without interruption. The aggregate architecture has been building, room by room, since the first tree established itself on that ground. The soil in a mature, undisturbed forest doesn't just have more organic matter than a managed field. It has more time. It has the compounded biological investment of every organism that ever lived and died in that system, every root that ever exhaled carbon into the dark, every fungal thread that ever produced glomalin and then broke down and left its mortar behind. That soil is the physical record of every ancestor that ever contributed to it. And it's richer for every one of them.

Your managed land is building that same record. Slowly, season by season, aggregate by aggregate. Every act of stewardship you make is a deposit into an account that compounds across generations. Every act of extraction or disruption is a withdrawal that costs not just you but every future steward who inherits what you leave behind. The Non-Local Legacy isn't an abstract concept. It's the most concrete financial reality in agriculture. The soil you hand forward is either richer or poorer than the soil you received, and the difference between those two outcomes is the sum of every management decision you made during your time on this land.

There's a specific, sensory way to understand this that goes deeper than any explanation. Find the oldest, most undisturbed patch of ground on your property.

The corner of a paddock that the machinery never quite reaches. The base of an old hedgerow. The strip along the creek that's been under permanent vegetation for as long as anyone can remember. Dig down six inches in that spot and hold the soil in your hands. Feel its weight. Smell it. Press it between your fingers and notice how it holds together, how it crumbles into irregular, dark, crumbly aggregates rather than smearing into a paste. Now walk to your most intensively managed field and do the same thing. The difference between what you hold in your two hands is the difference between a Non-Local Legacy and a Non-Local Debt. One is the gift of time and undisturbed biology. The other is the cost of the Mechanical Metronome running without a biological clock to answer to.

You're building one of those two things right now. Every season. Every decision. Every time you choose whether to run the night tillage pass or leave the soil alone to breathe. The Non-Local Legacy isn't something you plan for retirement. It's something you're creating or destroying today, in real time, with every action you take or choose not to take on your land.

The practical implication is specific. Walk your land this week with one question in mind: what am I building here that will still be here in fifty years? Not the crop. Not the fence. The biological infrastructure. The fungal network. The aggregate architecture. The hedgerow. The perennial strip along the waterway. The slow-growing trees you planted two seasons ago that are only knee-high right now. These are your Non-Local Legacy assets. They're the terms in the sequence that will compound longest. Identify them. Protect them. Add to them deliberately, at least once per season, in a way that extends their compounding horizon beyond your own.

The Fibonacci sequence has no end. Each term is larger than the last. Each term carries the memory of everything that came before it. Your stewardship is a term in that sequence. Make it one that the terms that follow can build on.

The Seventh Generation Protocol

There's an old principle from Indigenous governance traditions that asks decision-makers to consider the impact of their choices on the seventh generation to come. Not the next quarter. Not the next election cycle. The seventh generation. Roughly 175 years into the future. The idea is that any decision worth making should be able to withstand that scrutiny. If it strengthens the world for people who don't exist yet, it's a good decision. If it weakens it, no short-term benefit justifies it.

The Seventh Generation Protocol, applied to soil stewardship through the lens of the Fibonacci Pulse, becomes a specific decision-making tool. Not a vague aspiration. A concrete filter that you can run every management decision through before you make it. The question is simple: does this action strengthen the Fibonacci Pulse for those I will never see?

Run that question against the five SRS horizons and it becomes immediately practical.

Does this tillage pass strengthen the Architecture for those I will never see? If the answer is no, because it shatters aggregates and severs fungal threads that took two seasons to rebuild, then the tillage pass fails the Protocol. Does this soluble phosphorus application strengthen the Biology for those I will never see? If the answer is no, because it suppresses the mycorrhizal trade that should be building the self-sustaining mineral delivery system, then the application fails the Protocol. Does leaving this field bare through winter strengthen the Armor for those I will never see? No. Does planting this slow-growing hardwood strip along the creek strengthen the Diversity and the Mycelial Internet for those I will never see? Yes, emphatically, because the root system of those trees will be weaving fungal connections and mining minerals and building glomalin for a century after you're gone.

The Protocol doesn't require perfection. It requires direction. You don't have to get every decision right. You have to get the direction right. Every season, the net sum of your decisions should be moving the SRS upward, the fungal network denser, the armor thicker, the architecture more open, the biological chord richer. That's the Seventh Generation Protocol in practice: a consistent directional commitment to building more than you extract, season after season, year after year, until the compounding does what compounding always does and produces something larger than any individual contribution could explain.

The Soil Resonance Score is the mathematical instrument of the Seventh Generation Protocol. It's how you translate the abstract commitment to future generations into a measurable, trackable, season-by-season trajectory. When you calculate your SRS at the start of each season and compare it to the same time last year, you're not just measuring soil health. You're measuring whether your stewardship is building a legacy or eroding one. An SRS that's climbing is a Non-Local Legacy in progress. An SRS that's flat or declining is a Non-Local Debt accumulating.

Here's a specific way to apply the Seventh Generation Protocol to your annual planning cycle. At the start of each season, before you plan any inputs or operations, take your current SRS scores for each horizon. For each horizon scoring below 6, ask the Seventh Generation question: what single action this season would most strengthen this horizon for those I will never see? Not the action that produces the fastest visible result. The action that builds the most durable biological infrastructure. For Architecture below 6, the answer is almost always the same: reduce mechanical disturbance and increase continuous living cover. For Armor below 6: increase compressed residue depth to five centimeters before the next extreme season. For Diversity below 6: add the missing functional role to the cover crop mix. For Biology below 6: identify what's suppressing the biological community and stop doing it. For Chemistry above 3: ask whether this input is building the biological delivery system or substituting for it, and if it's substituting, reduce the rate and redirect the investment toward the horizon that's causing the deficiency.

Write the answers down. These become your Seventh Generation commitments for the season. Not your full management plan. Your biological legacy commitments. The decisions that are specifically made for the people who will stand on this land long after you're gone. Everything else in your management plan can be driven by economics, by logistics, by weather. These specific commitments are driven by the Protocol. They're non-negotiable regardless of what the season throws at you, because they're the terms in the sequence that compound longest and matter most to the steward who inherits what you build.

Consider a hypothetical composite who captures this shift well. Picture someone like Mara, a 46-year-old grazier who took over her family's 300-acre property eight years ago. The land had been in continuous annual cropping for thirty years before she inherited it. The SRS when she started was roughly 35 out of a possible 190. The Architecture was a 2. The Armor was a 3. The Biology was a 2. She knew the practices. She'd read the books. But she kept making decisions based on the current season's economics rather than the seventh generation's inheritance. In year four, she sat down and wrote a Legacy Letter to her great-great-grandchildren. She described the land as it was when she received it and the land she intended to leave. That letter changed something. The following season, she made three decisions she couldn't economically justify in the short term: she planted a 200-meter riparian buffer of slow-growing natives along the creek, she stopped baling her cover crop residue and left it as a Brown Shield, and she designated a two-hectare sanctuary zone that no machinery would enter for at least ten years. Her accountant questioned all three. Her SRS, four seasons later, had climbed to 98. The riparian buffer's fungal network was already extending into the adjacent paddocks. The sanctuary zone showed Slake Test results that made visiting agronomists stop talking mid-sentence. Mara hadn't just improved her land. She had changed its trajectory. The seventh generation of stewards on that property will inherit a biological system that's already past the Inversion Point, compounding on its own, because one steward in one season asked the right question and committed to the answer.

The Seventh Generation Protocol is not a sacrifice. It's the highest-return investment strategy available in agriculture, because it's the only one that compounds across a timeline long enough for the Fibonacci sequence to fully express itself. The early terms are small. The ratio is constant. The sequence has no ceiling. And the steward who plants past their own horizon is the one who changes the trajectory of everything that comes after them.

The Return to the Source Code

The journey this book has traced began underground, in the invisible architecture of a soil aggregate held together by a glycoprotein discovered in 1996. It moved through the mathematics of a sequence that appears in sunflower heads and galaxy arms and the branching of your own blood vessels. It traveled through the night, listening to the earth breathe. It followed fungal threads through four hundred million years of underground commerce. It stood in fields at 2 AM feeling the warmth of a soil that was drinking the atmosphere. It looked at pests as tuning forks and weeds as diagnostic maps and bare ground as a biological emergency. It learned to see a conductor where there used to be a manager, an ancestor where there used to be a farmer, a cathedral where there used to be a pile of dirt.

That's the journey from the Linear Void to the Resonant Future. And it's not a journey you make once and then arrive. It's a spiral. Each season brings you back to the same questions, but from a higher vantage point. The Slake Test you run this autumn will show you something the Slake Test last autumn couldn't, because the soil has changed and so have you. The SRS you calculate next spring will carry the memory of every season of stewardship you've invested, compounding the way the sequence always compounds, each term larger than the last because it carries the sum of everything that came before it.

The Source Code of the soil is not complicated. It's recursive. It looks at what happened yesterday, adds it to what's happening today, and uses that sum to build tomorrow. $F_{n-1} + F_{n-2} = F_n$. The biology that your previous stewardship built, combined with the biology you're building now, produces a future that's mathematically more abundant than either alone. This is the formula the forest runs on. This is the formula the undisturbed prairie runs on. This is the formula your soil has been trying to run on since the first seed fell on it. Your job, the whole of your job, has always been to stop interrupting the sequence.

The peace that comes with this understanding is real and specific. It's not the peace of having solved every problem. It's the peace of knowing which problems are worth solving and which ones are the soil's own business. The drought isn't your enemy. It's a rhythm to be respected by a system with enough architectural resilience to breathe through it. The pest isn't your enemy. It's a tuning fork pointing at the SRS horizon that needs attention. The bare patch isn't a failure. It's an invitation to extend the biological chord into the frequency gap the soil is broadcasting. None of these things require you to fight. They require you to listen, to read the signal, and to respond with the specific biological action that the Fibonacci Pulse needs to keep compounding.

The shift from the Linear Void to the Resonant Future is a shift in identity as much as a shift in practice. The Extractor becomes the Steward. The Mechanic becomes the Alchemist. The Victim becomes the Observer. The Manager becomes the Conductor. And finally, at the deepest level, the Farmer becomes the Quantum Ancestor. Someone whose work doesn't end at the fence line or the property boundary or the edge of their own lifetime, but propagates forward through the entangled biological field of a living soil system, reaching people and places and seasons that no single human life could ever span.

You're not alone in a dying world. You're a term in the oldest recursive sequence on earth, connected backward to every steward who ever protected a pore space or refused to sever a fungal thread, and connected forward to every steward who will inherit what you build. The Fibonacci Pulse was running before you arrived. It will run long after you leave. Your contribution is the term you add while you're here. Make it a term that the sequence can build on.

The Source Code is activated. The pulse is beating. The cathedral has rooms. The mason is working in the dark. The Solar Pump is running. The Great Weaver is trading sugar for stone. The armor is holding the gradient. The atmospheric draw is pulling the night's breath deep into the biological community that has been waiting for it for four hundred million years. All of it is running, right now, beneath your feet, on the land you're responsible for, compounding toward a biological abundance that your great-great-grandchildren will stand inside and breathe deeply and never fully understand the origin of.

That origin is you. That origin is now. That origin is every specific, deliberate, biologically informed decision you make from this day forward, each one a handshake across time, each one a term in the sequence, each one a room added to the cathedral that the seventh generation will stand inside and call home.

Chapter Recap and Action Steps

This final chapter asks for four specific actions. Not suggestions. Commitments. Each one is designed to extend your stewardship beyond your own horizon and leave a biological legacy that compounds forward through the Fibonacci sequence long after your direct involvement ends.

The first action is to write your Land Covenant. This is a one-to-two-page document, written for whoever stewards this land after you, that describes what you built, why you built it, and what they'll need to know to keep it building. Write the current SRS scores for each horizon. Describe the sanctuary zone you've established and why it matters. Name the perennial species you've planted and where they are on the property. Explain the Slake Test and what a passing result looks like on this specific soil type. Describe the cover crop mix that's been working and the species that haven't. Write it plainly, the way you'd write it for someone who loves the land but doesn't yet know the Fibonacci Pulse. Date it. Keep one copy in the farmhouse and one in a waterproof container somewhere on the property itself. This document is your biological handshake across time. It's the instruction manual for the cathedral you've been building. Without it, the next steward starts from scratch. With it, they start from where you left off, carrying your term in the sequence forward into theirs.

The second action is to establish a permanent Sanctuary Zone. Choose an area of at least half a hectare, ideally one that's representative of your dominant soil type and not your worst corner. Mark its boundaries clearly with permanent markers. Commit in writing, in your farm records and in your Land Covenant, that no tillage implement will enter this zone, no synthetic input will be applied within it, and no machinery heavier than a person on foot will cross its surface for a minimum of ten years. Seed it with a diverse, perennial-dominated mix appropriate to your region: native grasses, deep-rooted forbs, and at least two nitrogen-fixing species. Let it establish without intervention. Run the Slake Test on it every spring. Count the fungal threads in its aggregates every autumn. Track its SRS trajectory separately from your managed paddocks. By year five, this zone will be your most persuasive argument for what the Fibonacci Pulse produces when the Mechanical Metronome is permanently switched off. By year ten, it will be the biological benchmark against which everything else on your property is measured. It will be the proof, in living soil, that the Source Code works.

The third action is to share the Fibonacci Pulse framework with one other person in your community before the next season begins. Not a lecture. A conversation. Show them your Slake Test results, before and after. Walk them through your SRS scores and what each number means. Let them hold a clod of your sanctuary zone soil in one hand and a clod from your most managed paddock in the other and feel the difference. Explain the Seventh Generation Protocol in your own words, the way it changed the decisions you make and the horizon you make them against. The Fibonacci Pulse compounds fastest when it spreads. Every steward who activates the Source Code on their land is adding a node to the biological network of the Resonant Future. Every conversation you have about this framework is a potential activation. The seventh generation doesn't just need your land to be a cathedral. They need a landscape of cathedrals. That landscape starts with the conversations you're willing to have now, before the results are dramatic enough to be undeniable, when the only evidence you have is the warmth of an armored soil at 2 AM and the thread of a fungal hypha visible in a broken aggregate and the stubborn integrity of a clod that refused to dissolve in a jar of water.

The fourth action is to calculate your current SRS one final time, then set a five-year trajectory target for each horizon. Not a single-season target. Five years. Write down where each horizon is today and where it needs to be in five years for your land to be past the Inversion Point and running primarily on its own biological intelligence. For most practitioners, that means Architecture above 7, Armor above 7, Diversity above 7, Biology above 6, and Chemistry below 4, because a Chemistry score below 4 in a biologically functional system means the biological delivery systems are doing the work that chemistry was previously substituting for. Calculate what your total SRS would be at those target scores: $(8\times7) + (5\times7) + (3\times7) + (2\times6) + (1\times3) = 56 + 35 + 21 + 12 + 3 = 127$ out of 190. That's 67%. That's past the Inversion Point. That's a soil system that's doing more work than you are, compounding on its own, building the Non-Local Legacy that the seventh generation will inherit. Write that number down. Write today's total next to it. The difference between those two numbers is the biological distance between where you are and where the Fibonacci Pulse can take you in five seasons of deliberate, Protocol-driven stewardship. Then put the paper somewhere you'll see it. Not to create pressure. To create direction. The sequence knows where it's going. Your job is to keep adding terms.

The pulse was always there. Under the compaction, under the chemistry, under the bare brown fields of the Linear Void, under every season of the Mechanical Metronome running without a biological clock to answer to. It was always there, waiting. Not patiently, because patience implies time, and the Fibonacci Pulse exists outside of time the way a mathematical truth exists outside of time. It simply is. It was always running. You were always part of it. The only thing that's changed is that now you know it. And knowing it, you can stop interrupting it. You can protect the pore. You can feed the Weaver. You can hold the gradient. You can plant past your own horizon. You can write the covenant and establish the sanctuary and have the conversation and set the five-year target and show up on your land every morning as the Conductor of a sequence that started four hundred million years ago and has no intention of stopping.

That's the whole job. That's the Fibonacci Pulse, activated. That's the handshake across time.

Conclusion: The Pulse Restored

The Ancestor Speaks: "My fully awakened one, my heart is full. The journey is complete, yet it is only just begun. We stand now, not at the edge of a broken field, but within the vibrant heart of a living Earth, under the boundless sky. Close your eyes. Can you feel the shift within you? The deep sigh of relief, the unfurling of understanding, the quiet, powerful knowing that has settled into your bones?

"You came to this book carrying the weariness of the **Linear Void**, a dissonance born of battling a machine that was never truly there. They told you the soil was dead, an empty ledger, and your heart echoed with the pain of that separation. But you dared to listen. You learned to see the **Silent Architecture**, not as inert clay, but as a breathing cathedral. You heard the ancient **Math of Grace**, the **Fibonacci Pulse**, humming from the heart of every sunflower, every galaxy, every strand of your own being. You witnessed the **Great Weaver**, tirelessly trading sugar for stone, and felt the immense sorrow of its severed threads.

"You learned to read the night, to honour the **Circadian Inhale**, a continuous, abundant blessing of free nitrogen and internal rain. You felt the Earth's skin, the **Living Armor**, shielding the **Metabolic Sweet Spot**, protecting the life beneath from the sun's harsh gaze and the raindrop's hammer. You saw the joyous **Carbon Exhale**, the sun's liquid light, feeding the dark earth, weaving connections, building a non-local legacy far beyond your own years. You stood as the **Quantum Observer**, witnessing the power of your own intention, collapsing the wave towards the cathedral, not the ruin. You moved from manager to **Conductor**, letting the Earth's own symphony play.

"And now, my child, you know the deepest truth: the **Fibonacci Pulse** was always there. Beneath the compaction, through the chemicals, under the bare, brown fields of forgetting. It never stopped. It was always running, in the undisturbed corner, in the ancient hedgerow, in the wild places where the **Mechanical Metronome** could not reach. It was always waiting, patiently, for you to remember its rhythm, for you to step out of its way.

"You are no longer an isolated atom in a dying world. You are a **Quantum Ancestor**, a living, breathing term in the oldest, most profound sequence on Earth. Every choice you make, every pore you protect, every fungal thread you refuse to sever, every tree you plant past your own horizon, is a **Handshake Across Time**. It is a direct, physical, undeniable message to the **Seventh Generation**, a promise that the world they inherit will be richer, more resilient, more alive, because you chose to listen.

"There will still be challenges. The drought may return. The markets will shift. But your relationship to these rhythms has changed. The drought is no longer an enemy, but a test of the architecture you have built. The pest is no longer a curse, but a tuning fork, guiding your hand. The bare patch is an invitation to extend the chorus, to sing a richer song. You are not fighting the land. You are dancing with it, moving to the same ancient, elegant music.

"Now, open your eyes. Look upon the world, and upon yourself, with this new sight, this profound knowing. Feel the deep, true peace that settles into your heart when you cease to fight and begin to conduct. The pulse is beating within you, and through you, resonating with the very heartbeat of the Earth. You are the architect, the weaver, the observer, the conductor, the ancestor. This is not a role you play. It is who you are.

"Go forth, my beloved. The Source Code is activated. The pulse is restored. The Earth is waiting for your frequency. Step into the Spiral and let your life be the most beautiful song."

* * *

* * *

* * *

* * *

* * *

The Pulse Restored

The spiral turns complete, its resonance now deep.

You are the ancestor, this wisdom you will keep.

Go forth, and walk as Master, of Earth's green, vibrant skin.

The Pulse is fully roused. The living world begins.

Conclusion

The Pulse Restored

You started this book standing at the edge of a field that wasn't working. The soil cracked in July. The inputs kept arriving and the returns kept shrinking. You were doing everything the manuals said, and the system was still fighting you. That feeling had a name. It was the Linear Void. And now, having walked through fifteen chapters of biology, mathematics, physics, and ancestral responsibility, you can see it for exactly what it was: a frequency mismatch between a living, recursive system and the flat, additive thinking that was being imposed on it.

That mismatch is over the moment you choose to end it.

The Silent Architecture is not a metaphor. It's a measurable, testable, scoreable reality that exists in every cubic centimeter of soil that has been left alone long enough for glomalin to do its work. The cathedral is real. The pore space is real. The rooms that hold water, circulate air, and house billions of organisms doing their sacred and specific work are real, physical structures that you can assess with a jar of water and a clod of earth the size of a lemon. When that clod holds in the water, the cathedral is standing. When it dissolves, you know exactly what to do next.

The Math of Grace is equally real. The Fibonacci weights in your Soil Resonance Score aren't decorative. They reflect the actual compounding hierarchy of a living soil system. Architecture at 8. Armor at 5. Diversity at 3. Biology at 2. Chemistry at 1. That sequence tells you where to invest your energy, in what order, and why. A soil with a strong foundation compounds its improvements. A soil with a weak foundation multiplies nothing, regardless of how much diversity or biology you add above it. The formula is the instruction manual for the sequence. The sequence is the operating code of every living system on earth. And now it's in your hands, weighted correctly, ready to be applied to the specific land beneath your feet.

The Mycelial Internet is four hundred million years old and it's been running on your property since before you arrived. The Great Weaver doesn't need your help. It needs your restraint. Every meter of hyphal thread you protect by leaving the soil undisturbed is a meter of compounding biological infrastructure that will produce more next season than it produced this one. Every tillage pass that severs it is a reset of the sequence back toward zero. That's not a judgment. It's arithmetic. And now you can do the arithmetic yourself.

The Circadian Inhale is happening tonight. Right now, as you read this, the soil on your land is cooling, the pressure gradients are shifting, and the atmospheric draw is pulling nitrogen, oxygen, and condensing moisture downward through every open pore channel in the profile. The nitrogen-fixing bacteria are at peak activity. The Carbon Anchor is running. The Internal Rain is depositing two to four millimeters of water equivalent into the aggregate cistern without a pump, without a bill, without any intervention from you. The only question is whether the pores are open enough to receive it and the armor is thick enough to hold the gradient that drives it.

The Quantum Observer Effect is not mysticism. It's the practical consequence of the fact that every field decision you make is filtered through the lens you use to see the field. A steward who sees bare soil as neutral makes different decisions than a steward who sees it as an emergency. A steward who reads a pest event as a failure makes different decisions than one who reads it as a tuning fork pointing at the SRS horizon that's out of tune. The lens you choose is the most consequential management decision you make all year. And it costs nothing to change.

All of it, the architecture, the math, the breath, the web, the armor, the carbon, the quantum observer, the ancestral responsibility, all of it is one thing. It's the Fibonacci Pulse. It's the Source Code of the soil, expressing itself through every system that has ever been left alone long enough to run its own mathematics. You didn't build it. You can't build it. You can only stop interrupting it and give it the conditions it needs to compound on its own terms, in its own time, toward a biological abundance that no linear model ever predicted and no mechanical metronome ever allowed.

Your Role as the Ancestor

The steward who finishes this book and goes back to managing a farm the same way they managed it before hasn't really finished this book. They've read it. Reading and finishing are different things. Finishing means the understanding has landed somewhere deep enough to change what you do on Monday morning, and the Monday after that, and the Monday of the season you're building for a child whose name you'll never know.

You are already a Quantum Ancestor. The carbon you anchor today is entangled with roots that haven't been planted yet. The glomalin you protect is building the cathedral that your great-grandchildren will stand inside. The fungal network you refuse to sever is weaving connections that will still be conducting mineral trades in soil that won't be yours to farm. These aren't motivational statements. They're the physical consequences of biological entanglement operating across time the same way it operates across space. Non-locally. Instantly. Without diminishing with distance.

The seventh generation is already in the probability field of your land. They're waiting for you to collapse the wave toward the cathedral rather than the ruin. Every management decision you make from this point forward is a vote on which of those two states gets expressed as physical reality. The tillage pass you don't run is a pore space that the Weaver will colonize this winter. The residue you don't bale is the Fungal Bridge that keeps the network alive through the cold months. The tree you plant past your own harvest horizon is the Solar Pump that will run 365 days a year for a century, depositing carbon into the aggregate architecture long after you're gone. These aren't sacrifices. They're the highest-return investments available in agriculture, because they're the only ones that compound across a timeline long enough for the Fibonacci sequence to fully express itself.

Step into your fields tomorrow not as a manager checking for problems, but as a Conductor who knows the score. The orchestra is already there. The strings are the microbes. The brass is the fungal network. The rhythm section is the Circadian Inhale, keeping the tempo through every season. Your job is to hold the space for the music, not to produce it yourself. Lower the baton when the system is playing well. Intervene when a section is out of tune, specifically, surgically, with the precision of a conductor who knows which instrument is off rather than the desperation of a mechanic who replaces everything hoping to find the fault.

The drought that's coming won't be your enemy. It will be a test of your architecture. The pest that arrives next season won't be a failure. It will be a tuning fork pointing at the SRS horizon that needs attention. The bare patch in the corner of the paddock won't be a problem. It will be an invitation to extend the biological chord into the frequency gap the soil is broadcasting. None of these things require you to fight. They require you to listen, to read the signal, and to respond with the specific biological action that the Fibonacci Pulse needs to keep compounding.

The resilience against drought you've been reading about isn't a product you buy. It's the aggregate cistern filling with Internal Rain every night, the Memory of the Rain stored in glomalin-coated pores that hold ten times their weight in water. The end of chemical dependence isn't a goal you set. It's the mathematical consequence of an SRS Architecture score above 7, where the biological delivery systems are dense enough to mine minerals from parent material without any help from a bag. The deep peace of being in tune isn't a feeling you manufacture. It's what arrives when the system is finally running its own source code and you're no longer fighting the frequency it runs on.

These outcomes are waiting. They're not waiting for a perfect season or a perfect farm or a perfect understanding of quantum biology. They're waiting for you to take the first specific, measurable, biological step in the direction of the pulse. And you know exactly what that step is, because you've been building toward it since Chapter 1.

The Step-by-Step Path to Resonance

Everything in this book converges into a sequence of five actions. Not a checklist. A spiral. Each action builds on the one before it, the way the Fibonacci sequence builds on itself, each term carrying the memory of everything that came before. Follow the spiral and the system compounds. Skip a term and the compounding stalls. The sequence matters as much as the direction.

The first action is the Slake Test. Do it today, or this week, before any other intervention. Pull a clod from three representative locations on your land. Let them air-dry for a full day. Drop them into clear jars of water and watch what happens at the thirty-second mark and the five-minute mark. Photograph both moments. Write down the location, the management history, and how long each clod held or how quickly it dissolved. This is your architectural baseline. This is your SRS Architecture score made visible. Everything else in the sequence starts here, because Architecture carries a weight of 8 and multiplies the value of every other investment you make. A clod that dissolves in thirty seconds is telling you that your first and only priority for the next twelve months is protecting the pore. A clod that holds for five minutes is telling you that the foundation is solid enough to start building the upper floors.

The second action is to calculate your initial Soil Resonance Score using the formula from Chapter 2. Score each of the five horizons honestly, from 0 to 10, using the specific criteria you've been given throughout this book. Architecture from your Slake Test. Armor from your compressed residue depth and surface coverage. Diversity from your species count and functional role coverage. Biology from your earthworm count and fungal thread score. Chemistry from how much of your nutrient supply is biological versus purchased. Multiply each score by its

Fibonacci weight: 8, 5, 3, 2, 1. Add them together. Write the total down with today's date. This number is your starting point. Not a grade. Not a judgment. A coordinate. You're here. The Inversion Point is at roughly 55 to 65% of the maximum possible score of 190, which means somewhere between 105 and 124. The distance between your current score and that threshold is the biological distance you're covering. Knowing it precisely is what allows you to track the compounding rather than just hoping for it.

The third action is to commit to 100% Armor from this day forward. Not 80%. Not mostly covered. One hundred percent. If you can see bare soil larger than your hand anywhere in your managed fields, you have a gap in the skin of the cathedral, and every day that gap exists, the glomalin is degrading, the gradient is collapsing, and the Circadian Inhale is weakening. This commitment doesn't require a new input. It requires a new default. The default changes from "bare is neutral" to "bare is emergency." Every harvest is followed within fourteen days by a cover crop establishment. Every gap in the cover crop canopy is addressed with a fast-establishing bridge species before the next extreme season arrives. Every baling decision is evaluated against the question of whether the residue is worth more as a Brown Shield than as a commodity. Most of the time, it is. The compressed residue depth target is five centimeters. Measure it. Hold it. The biology inside it will do the rest.

The fourth action is to introduce the Rule of Five in your next cover crop planting. Not five random species. Five functional roles: one Nitrogen Fixer, one Deep Rooter, one Carbon Builder, one Pollinator Bridge, and one Mycorrhizal Host. The specific species matter less than the functional completeness. Fill all five roles and the biological chord has enough voices to start self-regulating. The Noise-Canceling effect kicks in. The Stress-Management Protocol of the Great Weaver activates. The Vertical Stack opens three floors of the soil profile simultaneously, feeding three distinct microbial communities, producing glomalin at three different depths, and building architectural complexity throughout the full biological zone. The Diversity score climbs. The Biology score follows. And because both of those horizons are now feeding the Architecture through glomalin production and biopore creation, the highest-weighted horizon in your SRS starts to compound faster than any single-horizon investment could produce.

The fifth action is to practice Intentional Observation during the Circadian Inhale. Once a week, go out onto your land between 10 PM and 2 AM. Not to work. Not to check on anything specific. To observe. Press your palm flat on a patch of armored soil and feel whether it's warmer than the air above it, whether it feels slightly damp, whether something is happening beneath your hand that isn't happening on the bare patch twenty meters away. Look for the leading edge of the wave collapse: the single earthworm in a paddock that previously had none, the fungal thread visible in a broken aggregate, the Slake Test clod that held for three minutes when last season it dissolved in thirty seconds. Write down what you find. One observation per walk. These observations are not just data. They're the act of a Quantum Observer collapsing the wave toward the cathedral. They're the practice of a Conductor listening to the orchestra before raising the baton. They're the specific, daily discipline of a Resonant Human who has stopped looking for problems and started witnessing potential.

Follow this spiral and you will witness the mud of today transform into something that no linear model predicted and no mechanical metronome allowed. Not because you forced it. Because you stopped interrupting it. The soil doesn't need you to work harder. It needs you to work with it. And working with it means understanding the sequence it runs on, protecting the architecture it builds, feeding the network it weaves, holding the armor that keeps it alive, and planting past your own horizon into the biological future that the seventh generation is waiting to inherit.

The Fibonacci Pulse was always there. Under every compacted field, under every bare winter paddock, under every season of inputs substituting for a biological system that was trying to come back online. It was always there, running in the undisturbed corner by the fence, in the old hedgerow, in the creek bank that the machinery never reached. It was always there, waiting for a steward who understood the sequence well enough to get out of its way.

That steward is you. That time is now. The pulse is beating. The cathedral has rooms. The mason is at work in the dark.

The Source Code is activated.

Reader's Guide

***The Fibonacci Pulse* by Zen Knezevic**

How to Use This Book

The Fibonacci Pulse is not designed to be rushed or skimmed. It invites readers to slow down, observe patterns, and reconsider assumptions about soil, stewardship, and systems that compound over time.

Some readers may choose to move sequentially. Others may return to certain chapters repeatedly as their land, thinking, or circumstances change. Both approaches are valid.

This guide is intended to support:

- individual reflection
- group discussion
- practical application in the field
- deeper engagement with the book's central ideas

Key Themes to Hold While Reading

As you move through the book, keep these questions in mind:

- Where do living systems **compound**, and where are they being **reset**?

- What does **Soil Intelligence** look like in practice, not theory?
- How does **time**—not effort—shape outcomes in living systems?
- Where does management cross the line into **interruption**?
- What responsibilities extend beyond a single season or generation?

Section by Section Reflection Questions

Prologue & Introduction — *Seeing Differently*

1. The prologue suggests that most outcomes are shaped beneath the surface.
   Where in your own work or life do unseen, structures play a larger role than visible actions?
2. The idea that *one teaspoon of soil contains more living organisms than there are people on Earth* reframes scale and complexity.
   How does this change your assumptions about control, intervention, or efficiency?
3. What does the book mean by "working against the pulse"?
   Where might you already be doing this without realising it?

Soil Intelligence & Architecture — *Foundations Matter*

1. How does the concept of **Soil Intelligence** differ from viewing soil as a resource or substrate?
2. The book emphasises structure, pore space, and biological networks over chemistry alone.
   How does this challenge conventional approaches you've been taught or practiced?
3. Where in your land or system do you see signs of collapsed architecture—or of recovery?

Compounding vs Linear Thinking — *The Fibonacci Pulse*

1. The book contrasts linear inputs with recursive, compounding systems. Where in your current practices are you expecting linear responses from non linear systems?
2. How does the Fibonacci sequence function here as a *pattern*, rather than a formula?
3. Can you identify one area—soil, business, community, or personal life —where compounding is already occurring?

Interruption, Disturbance & the Linear Void

1. The concept of the "Linear Void" describes systems driven by short term efficiency and repeated disturbance.
   Where do you see this operating most strongly in modern agriculture or management?
2. The book argues that interruption carries hidden costs that appear later.
   What examples have you witnessed where this is true?
3. How does reframing problems as **signals** rather than failures change your response to them?

Stewardship, Time & the Seventh Generation

1. What does stewardship mean when outcomes extend beyond your own lifetime?
2. The book honours ancestors who "left more than they took."
   What would that look like in your specific context?
3. How does thinking in terms of the **Seventh Generation** alter decision making today?

Practical Application Prompts

Use these to translate insight into action:

- Identify **one practice** that interrupts biological continuity.
  What would it take to reduce or remove that interruption?
- Choose **one indicator** of compounding (soil structure, water retention, biological diversity).
  How might you measure progress differently over time?
- Observe one place on your land that is *not actively managed.*
  What does it reveal about intelligence, resilience, or recovery?

Group Discussion Suggestions

For reading groups or facilitated sessions:

- Begin each session by asking:
  "What did this section help you see differently?"
- Encourage participants to share **observations**, not just opinions.
- Allow silence. This book benefits from space and reflection.
- Consider pairing discussion with a **field walk**, even a short one.

Closing Reflection

The Fibonacci Pulse ultimately asks a quiet but demanding question:

> Are we willing to slow down long enough to work with living systems rather than against them?

The answers will differ for every reader—but the responsibility is shared.

Appendix - Soil Resonance Score (SRS) — Field Worksheet

Soil Resonance Score (SRS) — Field Worksheet

Purpose:

This worksheet helps you measure what your land is *building*, not just what it is producing.

There are no "right" scores—only honest baselines.

Step 1 — Record Context

Date: ____________________________

Paddock / Field Name: ____________________________

Soil Type (if known): ____________________________

Recent Management (last 12 months):

Step 2 — Score Each Horizon (0–10)

Circle the score that best matches **what you can observe today**, not what you intend.

1. STRUCTURE (Weight ×8)

Aggregate stability, pore space, infiltration

0–1 Clod dissolves immediately; surface seals; frequent pooling

2–3 Clod breaks down fast; slow infiltration; crusting common

4–5 Partial structure; clod holds briefly; mixed results after rain

6–7 Good crumb; clod holds minutes; water enters soil

8–9 Strong aggregates; rapid infiltration; soil feels open

10 Exceptional structure; resilient in drought and deluge

Your score: ______ × 8 = ______

2. ARMOR (Weight ×5)

Surface protection: living plants and residue

0–1 Soil often bare; long brown gaps

2–3 Patchy cover; residue thin or inconsistent

4–5 Seasonal cover; noticeable gaps

6–7 Consistent cover most of the year

8–9 Dense green/brown cover; minimal bare soil

10 Continuous 365 day soil armor

Your score: ______ × 5 = ______

3. DIVERSITY (Weight ×3)

Functional diversity, not just species count

0–1 Monoculture or near monoculture

2–3 2–3 species; limited function

4–5 3–4 species; some functional range

6–7 5+ species; multiple roles present

8–9 Broad functional diversity; seasonal overlap

10 Redundant, resilient diversity across seasons

Your score: ______ × 3 = ______

---

4. LIFE (Weight ×2)

Visible soil biology

0–1 Few signs of life; little smell; poor crumb

2–3 Occasional worms; biology thin

4–5 Some worms; moderate activity

6–7 Worms common; soil smells alive

8–9 Strong fungal presence; rapid residue breakdown

10 Abundant, diverse soil life

Your score: ______ × 2 = ______

---

5. NUTRIENTS (Weight ×1)

How fertility is delivered

0–1 Fully dependent on purchased inputs

2–3 High supplementation; little biological supply

4–5 Mixed system; partial biological cycling

6–7 Mostly biological; inputs declining

8–9 Nutrients largely cycled biologically

10 Nutrients delivered almost entirely by biology

Your score: ______ × 1 = ______

Step 3 — Calculate Your SRS

Add all weighted scores:

STRUCTURE: ______

ARMOR: ______

DIVERSITY: ______

LIFE: ______

NUTRIENTS: ______

SRS BASELINE TOTAL:

_______ / 190

Step 4 — Interpret (Not Judge)

Use this only as orientation:

- **0–50** Foundation rebuilding phase
- **51–100** Compounding beginning
- **101–150** System doing more work itself
- **151–190** Long established soil intelligence

(These are descriptors, not grades.)

Step 5 — Choose One High Leverage Focus

Which **highest weight horizon** is lowest?

Structure (×8)

Armor (×5)

One action I will take in the next 14 days to improve it:

Step 6 — Commit to Observation

Next SRS check date (3–6 months): ______________________________

Small changes at high weight layers compound faster than big changes at low weight layers.

Acknowledgments

This book did not come into being alone.

It grew the way living things grow — through relationship, through patience, and through the quiet contributions of many hand's, some seen, many unseen.

I owe a deep debt of gratitude to those who walked beside this work while it was still forming. To **Norman Olsen**, whose careful and steady insight helped this manuscript find its footing — your questions sharpened what mattered, and your feedback arrived not as correction, but as guidance. Thank you for listening closely to what the soil itself was trying to say.

To my family, whose patience was tested more times than can be counted — thank you for holding space while I wandered into ideas that could not be rushed back. Thank you for the silence you allowed, the time you gave, and the faith you kept when the work asked for more than was convenient. No book like this is written without borrowing heavily from those who love the author, and I acknowledge that debt with humility.

There are also acknowledgements that do not fit neatly into names.

This work was shaped by ancestors I never met — farmers, stewards, and walkers of land whose hands broke branches while climbing fences, whose boots compacted soil without knowing what lived beneath, whose tools tore unseen threads long before those threads had names. I carry both their harm and their inheritance. I write this book not in judgment of them, but in conversation with them, aware that understanding often arrives after action, and responsibility begins when awareness does.

Because the subject of this book deals with invisible systems, subterranean intelligence, and processes that resist literal illustration, the cover image was generated using artificial intelligence to better convey the underlying concept. The vision itself was clear — to reflect soil not as dirt, but as living architecture and memory — and the tool was chosen not as a shortcut, but as a translator for something that exists beneath language and beyond the surface.

And then there is **Lucky**.

Lucky, who does not care about deadlines, arguments, or elegant metaphors. Lucky, who believes with complete certainty that no idea is more important than a walk, no paragraph more urgent than a thrown stick, and no long stretch of writing should go uninterrupted. This book was paused more times than I can count at the insistence of a cold nose on my knee, a paw on the keyboard, or a stare that made it clear I had wandered too far into abstraction and needed to return to the living world. I obeyed him often. This work is better for it.

Finally, I acknowledge the living systems themselves.

The soil that held while it was misunderstood.
The fungi that rebuilt after being cut.
The organisms crushed beneath careless steps — and those that returned anyway.
The quiet intelligence that continued its work long before we learned how to listen.

This book is offered as an act of attention — and as a promise to tread more gently now that I know what has always been there.

Made in the USA
Monee, IL
08 November 2026

67816262R00173